COMPREHENSIVE ANALYTICAL CHEMISTRY

ELSEVIER PUBLISHING COMPANY
335 JAN VAN GALENSTRAAT, P.O. BOX 211, AMSTERDAM

AMERICAN ELSEVIER PUBLISHING COMPANY, INC.
52 VANDERBILT AVENUE, NEW YORK 17, N.Y.

ELSEVIER PUBLISHING COMPANY LIMITED
12B RIPPLESIDE COMMERCIAL ESTATE
RIPPLE ROAD, BARKING, ESSEX

Library of Congress Catalog Card Number 58–10158
With 80 illustrations and 18 tables

PRINTED IN THE NETHERLANDS BY
DRUKKERIJ G.J.THIEME N.V., NIJMEGEN

COMPREHENSIVE ANALYTICAL CHEMISTRY

ADVISORY BOARD

COMPREHENSIVE ANALYTICAL CHEMISTRY

VOLUME I

CLASSICAL ANALYSIS

★

VOLUME II

ELECTRICAL METHODS
PHYSICAL SEPARATION METHODS

★

VOLUME III

OPTICAL METHODS

★

VOLUME IV

INDUSTRIAL AND
OTHER SPECIALIST APPLICATIONS

★

VOLUME V

MISCELLANEOUS METHODS
GENERAL INDEX

★

COMPREHENSIVE ANALYTICAL CHEMISTRY

Edited by

CECIL L. WILSON, PH.D., D.SC., F.R.I.C., F.I.C.I.

Professor of Inorganic and Analytical Chemistry
The Queen's University of Belfast

and

DAVID W. WILSON, M.SC., F.R.I.C.

Head of the Chemistry Department
Sir John Cass College, London

in association with

C. R. N. STROUTS, M.A., F.R.I.C.

Formerly Chairman of the Analytical Chemists' Committee
of Imperial Chemical Industries, Ltd.

VOLUME IIA

ELECTRICAL METHODS

ELSEVIER PUBLISHING COMPANY

AMSTERDAM LONDON NEW YORK

1964

Contributors to Volume II A

A. J. LINDSEY PH.D., D.SC., F.R.I.C., M.I.E.E.	Research Laboratories, Carreras Ltd., Wickford, Essex
DONALD G. DAVIS, JR. PH.D.	Professor of Chemistry, Louisiana State University, New Orleans, La.
T. S. BURKHALTER PH.D.	Director, Materials Research and Development Laboratory, Texas Instrument Incorporated, Dallas, Tex.

Preface

In *Comprehensive Analytical Chemistry* the aim is to provide a work which, in many instances, should be a self-sufficient reference work; but where this is not possible, it should at least be a starting point for any analytical investigation.

It is hoped to include the widest selection of analytical topics that is possible within the compass of the work; and to give material in sufficient detail to allow it to be utilised directly, not only by professional analytical chemists, but also by those workers whose use of analytical methods is incidental to their other work rather than continual. Where it is not possible to give details of methods, full reference to the pertinent original literature will be included.

The aim is, in other words, to present *Comprehensive Analytical Chemistry* as a *working* manual, offering as far as possible, direct and immediate assistance.

This first part of Volume II, dealing with some electrical methods of analysis, begins the presentation of the various instrumental and physical methods of analysis that have been developed, largely in recent years, to supplement the classical processes described in Volume I. Although some of the instrumental methods have historically a classical standing, the newer methods have extended the resources, and hence the capabilities, of the analytical chemist in a way that could scarcely have been envisaged even twenty years ago.

A decision has been made to publish this and subsequent Volumes in smaller parts. Volume I, which appeared in three parts, comprised in all more than 2000 pages. In spite of its most encouraging reception, there was the legitimate criticism that each part was inconveniently large.

This new policy will result in parts which are easier to handle, and will also, by giving the Editors greater flexibility in arranging the content, minimise delays in publication due to inevitable hazards in the preparation of individual chapters.

Once again the Editors would express their appreciation of help given by members of the Advisory Board, and of the friendly co-operation, both of the individual authors, and of the publishers, who have continued to be both long-suffering and understanding.

Special thanks are also owed to Dr. A. J. Lindsey, who has not only contributed one of the sections, but who has also undertaken the difficult task of writing an Introduction to this part. The Editors are also very grateful to Mr. C. R. N. Strouts, who has lightened their task appreciably by undertaking a considerable part of the sub-editing of manuscripts and proofs for the press.

Belfast and London, November 1963

C.L.W.
D.W.W.

Contents

VOLUME IIB ELECTRICAL METHODS *(continued)*

VOLUME IIC PHYSICAL METHODS

VOLUME IID PHYSICAL SEPARATION METHODS

VOLUME IIE PHYSICAL SEPARATION METHODS *(continued)*

Chapter I

Electrochemical Analysis: Introduction

ARTHUR J. LINDSEY

1. General

Since almost every type of chemical process involves some kind of charge transfer it could be argued that the whole of chemical analysis is electrochemical in nature. However, it is generally understood that an electrochemical analysis involves the use of an external electrical device capable of inducing charge transfers in a chemical system with the object of detecting or determining the constituents of the system. Within this very broad definition are included the well established methods of electrodeposition, potentiometry and conductimetry, and voltammetry including polarography, and a number of extensions of these methods and combinations of them with other established techniques. Thus, one of the most valuable contributions of electrochemistry to analysis has been its application to end-point detection in volumetric methods.

In this volume, analytical methods classified under the headings of electrodeposition, potentiometric titrations, conductometric titrations and high-frequency (impedimetric) titrations are described. Polarography, coulometry and related current–voltage methods are to be discussed in a later volume.

The bases of all these are the fundamental principles of the stoichiometry of ionic reactions, of the thermodynamics of electrode and of phase boundary equilibria and of the kinetics of motion of charged particles in media of varying viscosity.

2. Basic Studies

For a fuller study of the fundamentals of electrochemistry and electrochemical analysis, reference should be made to a number of electrochemical works, some of which are listed in the bibliography to Chapter II, page 61; a brief, but most lucid, discussion of electrochemical thermodynamics has been given by Ives and Janz[1]. A study of electrical theory and electronics and of their applications, especially in instrumentation and control, is recom-

References p. 6

mended as a background for the use of electrochemical methods of analysis, and should be regarded as an important preparation for these methods.

3. Units, Symbols and Conventions

Electrical units, of necessity employed in the diverse fields of physics, chemistry and engineering, have been developed over a long time and have become a matter of agreement between groups of scientific workers in a given field of study. Not always have the agreed conventions in one group corresponded exactly with those of another, and some confusion may be experienced by new workers in electrochemical analysis when consulting a number of sources of information of varying origin. Generally, divergencies are small in magnitude (although they could be great in fundamental conception), so that from a practical point of view for analytical work, as distinct from precision standardisation of electrical measuring equipment, no difficulties are likely to arise.

(a) Units

For many years the international system of units was adopted. In this, the units of current and of resistance were arbitrarily defined and that of voltage was a derived unit from these. All other units were also derived and their magnitudes were agreed by international co-operation in standardisation. The magnitude of these units differed only by small amounts (a few parts in 10,000) from absolute units derived from the centimetre, gram and second (c.g.s. system). The international system was discarded by agreement in 1948 and absolute units were adopted. More recently, in engineering and physical work, strong recommendation has been made for the m.k.s. system, based upon the metre, kilogram and second. Of course, this causes no change in the value of the working electrical units.

(b) Symbols

The physico-chemical quantities that are commonly represented by symbols are too numerous to be accommodated by ordinary printed alphabets and so each symbol has more than one significance. The conventions for symbols adopted by most publishing societies agree closely, and it is recommended that those listed by these societies should be employed in electrochemical analysis[2, 3].

As far as possible, these recommendations have been followed in the present volume.

(c) Sign Conventions

The variation of sign convention, from time to time and geographically, has been a serious source of confusion in electrochemistry and reference to past literature on the subject will inevitably call for a knowledge of the conventions that have been employed. A recommendation made by the International Union of Pure and Applied Chemistry[4], which has been called the Stockholm Convention, has now been generally adopted by electrochemists. The arguments in favour of this convention have been clearly discussed with illustrative examples by Ives and Janz[1], p. 26, and are abbreviated from these authors below.

(i) The e.m.f. of an electrochemical cell is related to the free energy change by the equation, $nFE = -\Delta G$.

(ii) The cell is represented diagrammatically so that its e.m.f. is the potential of the right-hand electrode measured with reference to the left-hand electrode.

(iii) The potential of an electrode is the e.m.f. of a cell formed from the electrode concerned on the right and a standard hydrogen electrode on the left.

In practice, the signs of electrode potentials on this system are the same as those formerly called "English" or "European" and are the reverse of those formerly called "American".

It is strongly recommended that the Stockholm Convention be adopted in all electroanalytical descriptions.

4. Classification of Electrochemical Methods of Analysis

Electrochemical techniques are most conveniently classified according to the processes employed in the method, and the means of attaining the analytical result from the method. The various techniques are reviewed below although only some of them are described in this volume.

(1) *Electrolytic Methods* include processes in which the separation of one or more constituents is effected by the passage of a current through a solution of the substance to be analysed. They may be quantitative or qualitative and on varying scales of amount, and include both electrodeposition and electrosolution methods.

(a) *Electrodeposition* techniques involve the deposition of metals upon cathodes, or of oxides upon anodes. Both can be quantitative and may thus be the basis of gravimetric electrodeposition methods. The quantitative deposition can however, be followed by re-solution followed by some other method of determination such as colorimetry. The use of counting procedures for radioactive deposits has given a quantitative method for sub-micro amounts of suitable elements. The development of mercury-cathode methods as a means of removing completely the more noble metals with the object of determining residual constituents (*e.g.* aluminium or magnesium) in the solution is a valuable technique. All of these can be transferred to the

References p. 6

microchemical scale (*i.e.* from a few decigrams to a few milligrams) without involving much elaboration of technique or precautions.

(b) *Internal electrolysis* (electrodeposition without an external source of current) has limited application but, for the determination of traces of more noble metals in the presence of much larger amounts of others, it is very satisfactory and easy to perform.

(c) *Electrosolution (electrographic) techniques,* involving anodic attack of the sample, are most useful methods of qualitative and rough quantitative examination of alloys and of larger objects that cannot be drilled or filed for sampling. It also has application in the examination of metal coatings for flaws.

In all the previously mentioned methods, electrical measurements of current or of potential are employed solely as means of control, although in other circumstances they can be used as a means of determination.

(2) *Potentiometry* is used in very few cases as a direct method of analysis; the most important of these is the determination of pH (hydrogen ion activity), but other isolated determinations of ion activity have been used as measures of concentration of constituents; for example, in the use of a silver chloride electrode to measure amounts of chloride ion.

The more common use of potentiometry is in end-point determinations for titrations. Here, the indicator electrode, usually in reversible equilibrium with one of the constituents, undergoes a considerable change in potential at the end point of the titration reaction. Polarised electrodes not in equilibrium with the electrolyte nevertheless represent a sufficient potential change at the end point to be satisfactory indicators.

Modern potential-measuring devices, especially those capable of operating in circuits of high impedance, cause no appreciable polarisation of the indicator electrode and thus are capable of giving a true representation of the potential of the electrode. Such devices are also capable of being the essential first stage in control mechanisms for other methods, as in potentiostats for control of electrodeposition, or as industrial process-controllers to provide automatic analytical control of such variables as pH or some other ion concentration. A similiar type of control is employed in automatic titration devices.

(3) *Amperometry,* determination by means of current measurement, is employed, especially in amperometric titrations, to give an indication of the end point. The method also depends upon potential control and is really a sub-division of the technique of *voltammetry* in which the reactant in an electrochemical process, selected by choice of potential which is then kept

constant, is measured by a diffusion current proportional to the concentration of the reactant. Voltammetry, the study of voltage–current relationships is, of course, another term for *polarography*, although the latter term is reserved by some authorities solely for techniques employing the dropping-mercury electrode.

(4) *Coulometry* has as its basis Faraday's laws of electrolysis, so that the total charge transfer gives a measure of the amount of material transferred. It involves the measurement of a constant current for a known time or of direct measurement of the total charge by means of an electromechanical or electrolytic coulometer. Coulometric titrations are specially adaptable to very small quantities in circumstances where ordinary titrations would be inappropriate or impossible, particularly where titrant is being consumed as it is generated so that no losses occur. An example is the electrogeneration of titanous ions.

(5) *Conductometry* is less frequently employed for direct determinations than as a method for end-point determinations in titrations. The basis of the method is the long-established law of Kohlrausch on independent migration of ions, and it involves the replacement of an ionic species by another of different conductance, the end point being determined by a change in direction (often to a minimum) of the conductance–volume curve.

(6) *Impedimetry*. Since only seldom can a circuit involving electrolytes be regarded as a pure resistance, *i.e.* with negligible reactance, most conductometric methods should be regarded as impedimetric. However, the term is more often applied to high-frequency methods in which no direct connection is made to the electrolyte. These titrimetric methods were originally called high-frequency conductance methods and they have the great advantage that the electrodes are usually plates placed outside the containing vessel. The end points are indicated by a change in impedance.

(7) *Polarography*, of all the methods listed, has the most complex theoretical basis, resting as it does upon diffusion characteristics of electrolytes and current–voltage relationships, as well as upon the physical nature and motion of the electrodes. In succession, a number of investigators have studied the limiting conditions at dropping-mercury and other micro electrodes and have related the concentration of the electro-active species to electrical characteristics of the system[5, 6, 7, 8, 9]. Very simple equipment can be used for manual production of diffusion-current/potential curves, but more elaborate automatic polarographs are quicker in use.

References p. 6

Modern polarographic techniques, which use electronic control and display of the current–voltage characteristics, are very rapid and accurate for the determination of trace quantities. They have the great advantage that a determination is repeatable in the same specimen many times over, either for the purpose of checking or to study a kinetic system undergoing rapid change.

Only with a knowledge of the theory of these techniques and of the possibilities and limitations of each can an effective choice be made for a particular analytical problem. The aim of this and other parts of volume II is to provide that knowledge.

References

1. D. J. G. Ives and G. J. Janz, *Reference Electrodes* (Academic Press, New York and London, 1961).
2. The Chemical Society, *Handbook for Chemical Society Authors* (The Chemical Society, London, 2nd. edn., 1961).
3. The Royal Society, *General Notes on the Preparation of Scientific Papers* (The Royal Society, London, 1950).
4. J. A. Christiansen and M. Pourbaix, *Compt. rend. conf. intern. chim. pure e appl.*, 17th Conf., Stockholm, p. 83 (1953).
5. D. Ilkovic, *Coll. Czech. Chem. Comm.*, 1934, **6**, 498.
6. J. J. Lingane, *J. Amer. Chem. Soc.*, 1953, **75**, 788.
7. P. Delahay *et al.*, *J. Amer. Chem. Soc.*, 1951, **73**, 4944, 5219.
8. H. A. Laitinen and I. M. Kolthoff, *J. Phys. Chem.*, 1941, **45**, 1079.
9. A. J. Lindsey, *J. Phys. Chem.*, 1952, **56**, 439.

Chapter II

Electrodeposition

ARTHUR J. LINDSEY

1. Introduction

Electrodeposition techniques include all analytical procedures in which a metal (but sometimes a compound such as lead dioxide) is deposited upon an electrode by the passage of an electric current. The process can often be quantitative and, in such cases, can usually be made the basis of gravimetric determination. However, the process could be an electrolytic separation prior to other methods of quantitative measurement. Qualitative deposition methods are included, as is also electrography which, as an electrosolution technique, is the reverse of electrodeposition.

Polarography and coulometry are excluded because, in these methods, the analytical results are computed from electrical measurements.

Electrodeposition methods as thus defined depend for their stoichiometry upon the well known laws of electrolysis first enunciated by Faraday, for their feasibility upon the thermodynamic relationships governing equilibrium at electrodes in a quiescent electrolyte, and for their speed of completion upon the laws of ion migration and diffusion in fluids. These fundamental considerations are therefore reviewed with special reference to their application in electrodeposition.

2. Theoretical Basis of Electrodeposition

Fundamentals of Electrochemistry

These matters are discussed only so far as they are essential to an understanding of the principles of analytical electrodeposition. A number of simplifications have been introduced; in particular, in the discussion of conductance where ion-interaction effects have been neglected. For further information, reference should be made to the standard works on theoretical electrochemistry[1,2,3,4].

References p. 61

(a) Electrolysis

When electrodes of conducting material are immersed in an electrolyte solution, ion migration occurs under the applied potential difference. The current passing through the solution is carried by the migration of cations (positively charged) to the cathode and of anions (negatively charged) to the anode. In order that a steady current may flow, the potential difference must be greater than a certain minimum value determined by the nature of the electrodes and the solution in which they are immersed. If the potential difference between an electrode and the solution is sufficient to cause a current to flow, the amount of electrochemical action (deposition of a metal, evolution of oxygen, *etc.*) is governed by the laws of electrolysis enunciated by Faraday. These are *(i)* "The amount of chemical decomposition or the amount of substance deposited on an electrode is proportional to the total charge passed through the electrolyte" and *(ii)* "The amounts of different substances liberated at an electrode by the same current flowing for the same time (the same total charge) are proportional to their chemical equivalents." The two laws may be summarized by the equation

$$m = \frac{IJt}{F}, \qquad (1)$$

where m = mass in grams of the liberated substance, J = equivalent weight, I = current in amperes, t = time for which the current flows in seconds, and F = Faraday's constant = 96,500 coulombs.

It should be noted that this relationship holds for the primary electrolytic process and that subsequent chemical action may follow to a varying extent. It is also possible that more than one electrolytic process may occur at the same time, for example, the deposition of a metal such as copper together with hydrogen. The value of m in equation (1) in such cases is the sum of the masses of all liberated species.

(b) Electrode Potential

An electrode of a metal immersed in a solution of its ions and in reversible equilibrium with them assumes a steady potential with respect to the solution. The potential is quantitatively expressed by the Nernst equation

$$E = E_0 + \frac{RT}{zF} \log_e a_m, \qquad (2)$$

where E = the potential difference in volts of the metal with respect to the solution and E_0 = the standard electrode potential in volts of the metal (in equilibrium with a solution of its ions at unit activity) with respect to a hydrogen electrode in reversible equilibrium with a solution of hydrogen ions at unit activity, z = the valency of the metal ions, a_m = the activity of the metal ions, R = the gas constant per mol, T = the absolute temperature and F = Faraday's constant. In Table II, 1, are given the standard electrode potentials corresponding to systems of interest to electrodeposition analysts.

TABLE II, 1

STANDARD POTENTIALS FOR ELECTRODE REACTIONS AT 25°

$M^{n+} + ne = M$

Ion	E_0 *Volts*	*Ion*	E_0 *Volts*	*Ion*	E_0 *Volts*
Au^{+}	+1.70	Cd^{2+}	−0.40	Lu^{3+}	−2.25
Au^{3+}	+1.50	Fe^{2+}	−0.44	Am^{3+}	−2.32
Pt^{2+}	+1.20	Ga^{3+}	−0.53	Y^{3+}	−2.37
Pd^{2+}	+0.99	Cr^{3+}	−0.71	Mg^{2+}	−2.37
Hg^{2+}	+0.86	Zn^{2+}	−0.76	La^{3+}	−2.52
Ag^{+}	+0.80	Nb^{3+}	−1.10	Gd^{3+}	−2.40
Hg_2^{2+}	+0.79	V^{2+}	−1.18	Sm^{3+}	−2.41
Cu^{+}	+0.52	Mn^{2+}	−1.18	Nd^{3+}	−2.44
Cu^{2+}	+0.34	Zr^{4+}	−1.53	Ce^{3+}	−2.48
Bi^{3+}	+0.20	Ti^{2+}	−1.63	Na^{+}	−2.71
Sb^{3+}	+0.10	Al^{3+}	−1.66	Ca^{2+}	−2.87
Sn^{4+}	+0.05	Hf^{4+}	−1.70	Sr^{2+}	−2.89
$2H^{+}$	±0.00	U^{3+}	−1.80	Ba^{2+}	−2.90
Pb^{2+}	−0.13	Be^{3+}	−1.85	Ra^{2+}	−2.92
Sn^{2+}	−0.14	Np^{3+}	−1.86	Cs^{+}	−2.93
Mo^{3+}	−0.20	Th^{4+}	−1.90	Rb^{+}	−2.93
Ni^{2+}	−0.25	Pu^{3+}	−2.07	K^{+}	−2.93
Co^{2+}	−0.28	Sc^{3+}	−2.08	Li^{+}	−3.05

A more practical form of this equation is

$$E = E_0 + \frac{0.0001982T}{z} \log_e a_m \qquad (3$$

or, for a univalent ion at room temperature (25°),

$$E = E_0 + 0.059 \log_{10} a_m. \qquad (4)$$

Since the activity of a metal ion species is approximately proportional to its concentration, it may be said that, at constant temperature,

$$\Delta E \propto \Delta \log_{10} C. \qquad (5)$$

The Nernst equation holds accurately only for a metal in reversible equilibrium with its ions.

References p. 61

(c) Electrodeposition

If the electrode is made more negative than the equilibrium value by connection to a source of current, the metal ions migrate to it and deposition will occur. If, on the other hand, the electrode is made more positive, ions will tend to dissolve from it. Electrodeposition upon metals other than that of the ionic species in solution will not follow these reversible criteria but, as soon as a coating of the deposited metal has been formed on the electrode, the potential will be governed by a similar relationship

$$E_D = E_0 - V + \frac{RT}{zF} \log_e a_m \tag{6}$$

or,

$$E_D = E_0 - V + 0.059 \log_{10} a_m \tag{7}$$

for a univalent ion at 25°. Here, E_D is the deposition potential, V is the over-voltage of the process and the other symbols have the significance previously defined. E_D is characteristic of the metal being deposited and is dependent upon the activity or the concentration of the metal in solution. E_D also changes with the current flowing in a manner illustrated in Fig. II, 1.

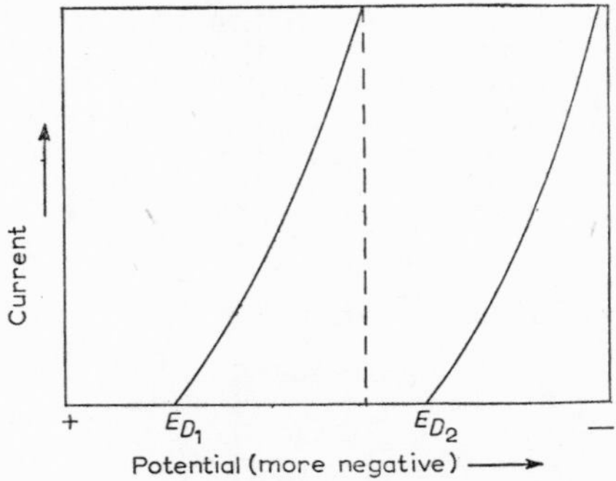

Fig. II, 1. Current-voltage relationship during electrodeposition.

Here, it will be seen that in a solution containing two metal ions, metal 1 will deposit under the conditions depicted and, even if metal 2 is present the latter will not reach its deposition potential when metal 1 is being deposited. However, if an attempt is made to deposit metal 1 completely, the last term in equation (7) will have an effect. When the activity (or concentration) has been reduced to one tenth of its original value, the deposition potential will become more negative by 0.059 volt so that the deposition curve will advance to the right by 0.059 volt. The same change will occur for each tenfold decrease in concentration. Thus, if quantitative results are required to one part in a thousand, the final deposition curve for metal 1 in Fig. II, 2 will be more negative than the initial deposition potential of metal 2 and both could be deposited together. If no deposition of the second metal is to occur, the electrode must not be made more negative than E_{D_2}. The final current, therefore, cannot be allowed to exceed the value I for a complete separation of metal 1 from metal 2.

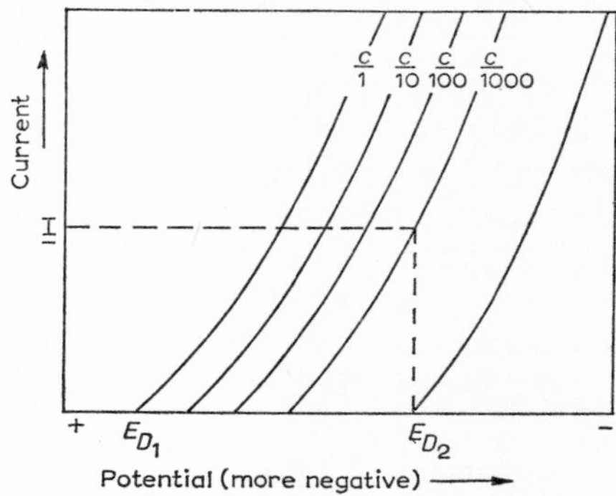

Fig. II, 2. Effect of concentration on electrodeposition curves.

These considerations are at the basis of quantitative separation of metals by electrodeposition using potential control.

(d) The Conductance of Solutions

The current-carrying capacity of all conductors depends upon the mobility of the charge-carrying particles. In metals, these are the valency electrons and their flow under the influence of a potential difference constitutes a negative current. The conventional direct electric current is said to flow in the opposite direction—the positive current—and this when flowing in a metallic conductor obeys Ohm's law. If an electrolytic cell is included in the circuit, by convention again the positive current is said to flow through the cell but, within the cell, the migration of all the charged particles (ions) under the influence of the applied potential difference makes a contribution towards the current. If a reference plane is considered in the centre of the electrolytic cell, the flow of all cations in one direction through this plane and of all anions in the other direction constitutes the current within the cell, and the total charge transfer rate is equal to the circuit current as indicated by an

References p. 61

external meter. For a pure electrolyte in solution, the ratio of the current carried by an ion to the total current is called the transport number of the ion. The contribution made to the total current by any single ionic species in an electrolyte solution is dependent upon the ionic mobility of the species as later defined.

The current-carrying capacity of an electrolytic solution (at a definite temperature) obeys Ohm's law,

$$I = \frac{E}{R} \qquad \text{or} \qquad C = \frac{1}{R} = \frac{I}{E}, \tag{8}$$

where I = current flowing in amperes, E = potential difference in volts between the ends of the body of electrolyte considered, R = resistance in ohms of the electrolyte, and C = conductance of the electrolyte or its current-carrying capacity. For a body of electrolyte of definite composition and temperature and of unit length and cross-section, C has a specific value K, the specific conductance or conductivity. Since in an electrolyte the ion migration provides the conductance, and since the charge carried by an ion depends upon its valency, it is useful for comparative purposes to list equivalent conductances.

The equivalent conductance of a pure electrolyte solution is defined by the equation

$$\Lambda = \mathrm{K}V, \tag{9}$$

where V is the volume containing one gram equivalent of the species under consideration. For increasing values of this volume, Λ approaches a limiting value Λ_∞, which corresponds to conditions in which all ions contribute their maximum effect towards the conductance. The effect of each ion is additive, so for a pure solution of a binary electrolyte

$$\Lambda_\infty = l_a + l_c, \tag{10}$$

where l_a = ionic conductance or mobility of the anion and l_c = ionic conductance or mobility of the cation. At a definite temperature the mobility of an ion has a definite value. The values for a number of common ions are given in Table II, 2.

Ionic mobilities increase with temperature and so larger currents can be carried through solutions when the temperature is raised.

(e) Diffusion

A further factor influencing the current flowing in an electrolyte, and therefore the rate of electrodeposition, is diffusion. When electrodeposition

TABLE II, 2

IONIC MOBILITIES AT 25°

Ion	l_a or l_c (*ohm*$^{-1}$ *cm*^{2})	*Ion*	l_a or l_c (*ohm*$^{-1}$ *cm*^{2})
H^+	350	$\frac{1}{2}Hg^{2+}$	63.6
OH^-	197.6	$\frac{1}{2}Ba^{2+}$	63.6
$\frac{1}{3}PO_4^{3-}$	92.8	$\frac{1}{3}Al^{3+}$	63.0
CN^-	82.0	$\frac{1}{2}Cu^{2+}$	59.5
$\frac{1}{2}SO_4^{2-}$	80.0	$\frac{1}{2}Sr^{2+}$	59.5
Rb^+	77.8	$\frac{1}{2}Cu^{2+}$	56.6
Tl^+	74.7	BrO_3^-	55.8
K^+	73.5	F^-	55.4
NH_4^+	73.4	$\frac{1}{2}Ni^{2+}$	53.5
NO_3^-	71.4	$\frac{1}{2}Mn^{2+}$	53.5
$\frac{1}{2}Pb^{2+}$	70.0	$\frac{1}{2}Mg^{2+}$	53.1
$\frac{1}{2}CO_3^{2-}$	69.3	$\frac{1}{2}Zn^{2+}$	52.8
$\frac{1}{2}Hg_2^{2+}$	68.6	CH_3COO^-	40.9
$\frac{1}{3}Fe^{3+}$	68.0	$C_6H_5COO^-$	32.3
ClO_4^-	67.3	$\frac{1}{2}C_2O_4^{2-}$	24.0
$\frac{1}{3}Cr^{3+}$	67.0		

commences, the solution in contact with the electrode is depleted in concentration, and ions from the bulk of the solution at the original concentration move toward the electrode under two forces, the electrostatic field causing migration and the thermal diffusion due to concentration differences. This diffusion gives rise to an activity gradient or concentration gradient near the electrode through the "diffusion layer". Diffusion through such a layer to a plane electrode is governed by Fick's law, originally expressed for concentration gradients,

$$\mathrm{d}N/\mathrm{d}t = AD\mathrm{d}a/\mathrm{d}x, \qquad (11)$$

where $\mathrm{d}N/\mathrm{d}t$ is the rate of diffusion of the species up to an area A of the plane electrode, $\mathrm{d}a/\mathrm{d}x$ is the activity gradient (assumed uniform) through the diffusion layer, and D is the diffusion coefficient.

In practice, the solutions used in deposition experiments are usually stirred vigorously, thus decreasing the thickness of the diffusion layer and increasing the gradient and hence the rate of diffusion. It is also common to introduce another electrolyte in greater concentration than the one to be deposited, and thus to carry more of the migration current by an ion not deposited under the conditions of the experiment; speed of deposition is thus further increased.

References p. 61

3. Quantitative Electrodeposition of Metals

The determination of metals by electrodeposition has been discussed in a number of textbooks [5–17] and reviews[18–23] and these sources of information, with earlier editions of the books, should provide a complete guide to the literature of the subject from earliest times.

(a) Classification of Metals for Electrodeposition

The metals may be conveniently divided into groups for electrodeposition analysis and this division is dependent upon the potential series of Table II, 1.

Several groupings are possible, but they all are arbitrary and one such grouping is shown in Table II, 3.

TABLE II, 3

GROUPING OF METALS INTO ELECTRODEPOSITION CLASSES

Group A, Precious Metal Group: Gold to Mercury.
Group B, Copper Group: Copper, Bismuth, Antimony, Arsenic.
Group C, Lead Group: Lead to Tin.
Group D, Zinc Group: Nickel to Zinc.
Group E, Mercury Cathode Group: Alkali metals, Lanthanons, *etc.*
Group F, Anodic Group: Metals anodically deposited as oxides.

Generally, metals in a group cannot be separated from each other by simple control of the applied potential.

(b) Conditions Governing the Formation of a Good Analytical Deposit

Three important conditions must be fulfilled, *(i)* deposition must be complete to the required degree of accuracy, *(ii)* the deposit must be in a state of purity (usually a deposited metal is pure, but it may occlude some material from the electrolyte), *(iii)* the deposit must be in suitable form, adhering well to the electrode and microcrystalline in form so as to withstand washing, drying and weighing without loss.

A trial experiment is a simple and sure way to check that the conditions of the analytical process are satisfactory, and such a procedure is recommended to ensure that the duration of the experiment and values of current, potential, temperature and stirring rate are all suitable. If hydrogen is liberated with the metal, the latter is often deposited in a spongy condition (a state that, in some cases, may be due to intermediate hydride formation). Such deposits tend to contain traces of electrolyte and therefore to weigh more than they should. Other reasons for excess of weight can be inaccurate potential control with consequent deposition of a second and unwanted metal, and weakly acidic solutions from which the metal can be formed with traces of oxide hydrolytically produced. High current densities can give rise to rough, heavily crystalline deposits, which adhere poorly to the electrode and thus cause low results.

(c) *The Effect of Hydrogen Overvoltage*

In general, hydrogen is not liberated at an electrode at a potential corresponding to that of a reversible hydrogen electrode in the same solution. With electrodes other than platinised platinum, a potential difference more negative than the reversible value is necessary to produce the liberation of gaseous hydrogen; the difference between this more negative potential and the reversible value is the "hydrogen overvoltage".

The hydrogen overvoltage depends upon the nature of the metal, its purity, the condition of its surface, the nature and concentration of the electrolyte, the current density and the temperature. In any arbitrarily chosen analytical system, it is therefore not possible to forecast the hydrogen overvoltage of the cathode. However, some values of minimum hydrogen overvoltage for various metal cathodes in normal sulphuric acid solution are given in Table II,4 to illustrate the range of variation that may be encountered, and the order in which the overvoltage on metals normally falls.

TABLE II, 4

MINIMUM HYDROGEN OVERVOLTAGES IN N SULPHURIC ACID

Metal	*Hydrogen Overvoltage*	*Metal*	*Hydrogen Overvoltage*
Platinum (platinised)	0.00	Copper	0.23
Gold	0.02	Cadmium	0.48
Platinum (bright)	0.09	Tin	0.53
Iron	0.14	Lead	0.64
Silver	0.15	Zinc	0.70
Nickel	0.21	Mercury	0.78

Towards the end of a determination by electrodeposition, the metal ion concentration diminishes to a very low value, and consequently the electrode potential becomes considerably more negative than its original value.

The hydrogen ion activity also increases, and both these factors increase the tendency for liberation of hydrogen towards the end of a determination. It is thus possible that the last traces of a metal may not be removed from a solution because all the current is utilised in hydrogen evolution.

References p. 61

(d) The Effect of Colloids in the Electrolyte

Certain colloidal constituents, such as gelatine, in the electrolyte may affect the character of a metal deposit. Generally, there is a tendency for colloids to produce a smoother and more compact coating, but sometimes the deposit is overweight because of gross adsorption of the colloidal material. The addition of colloidal substances, although common in electroplating techniques, is for this reason not used in analytical procedures.

(e) The Effect of Electrolyte Composition on Quantitative Deposition

(*i*) *The Effect of Anions*

The nature of the anion is governed by the choice of original sample and the mode of preparing a solution for analysis. Since electrodeposition methods are especially useful in analysis of alloys, the anions present are usually the result of the technique employed to produce a solution of the alloy. Hydrochloric acid solutions, often produced by dissolving the alloy in this acid with occasional additions of chlorate during dissolution, can be used for determinations of a large number of metals. Evolution of chlorine from the anode must be avoided since this causes attack of the anode, and deposition of its material (platinum) with the analytical deposit occurs. Chlorine evolution is prevented by use of suitable anodic depolarisers; hydrazine or hydroxylamine are the most suitable. Sulphuric acid solutions are not often used, although the presence of sulphuric acid with other anions is more common. Nitric acid solutions are satisfactory for many analyses, and in such solutions hydrogen evolution is inhibited, since the acid acts as a cathodic depolariser. Thus, copper and bismuth are often deposited in poor form when hydrogen is liberated at the same time, but from nitric acid solutions smooth quantitative deposits are produced. If the concentration of nitric acid is high, it may prevent cathodic deposition of the metal and, in some conditions, this is also desirable, for example, in the anodic deposition of lead as dioxide where cathodic deposition is to be avoided.

(*ii*) *The Effect of Acidity of the Solution*

Some effects have already been discussed under hydrogen overpotential. The metals in groups A, B and C are all deposited quantitatively in strongly acid solutions, but more precise control of acidity is necessary for groups D and E. Details are given under the techniques for individual metals.

(*iii*) *The Effect of Complex Ions*

The addition of compounds to form complex ions is often made to alter the relative deposition potentials of metals and thus to allow alternative separation procedures to be carried out. Thus, cupric ions have a deposition potential far less negative than that of cadmium ions, and copper is normally deposited quantitatively before any cadmium is deposited. On adding potassium cyanide, potassium cuprocyanide is formed, which is in equilibrium with only very low concentrations of cupric ions, and the cadmium can then be deposited quantitatively before copper begins to be deposited.

4. Apparatus for Electrodeposition Analysis

(a) Introduction

Cathodic deposition methods can be divided into two major groups: slow and rapid. The former were at one time often used, but were limited in application. Most often, a platinum dish was used as a cathode with a helical wire anode suspended in the electrolyte. It was customary to allow electrolysis to proceed without potential control for a prolonged period (often overnight). The method has limited application, but can be used for copper in brasses and other analyses in which no interference is possible from other metals.

The rapid technique has almost entirely superseded the slow method.

(b) Sources of Current

Currents up to ten amperes at low voltages only are often needed, and so lead accumulators are the best source of current. The most satisfactory arrangement is a car type of battery with a mains-operated charging equipment so that the battery may be kept fully charged during the time it is not in use. It is possible to use alternating current mains with a low-voltage transformer and a bridge metal rectifier with a suitable smoothing circuit; but the battery method is less expensive in initial cost. Regulation of current is best effected by means of a heavy wire-wound resistance with a sliding contact. The resistance should carry the necessary currents easily. An open-scale ammeter reading up to 10 amperes, and a voltmeter to check the battery, are also necessary. Some simple circuit arrangements are shown in Fig. II,3.

(c) Electrode Systems

The best material for electrodes is platinum, but it has the disadvantage that it is expensive. However, although the initial cost of platinum electrodes is high, the metal has an equally high recovery value when replacement is necessary. Its use is an economically sound proposition when compared with that of other electrode materials. The most suitable design of electrodes for general use is a pair of concentric gauze cylinders, the outer one of which is readily detachable from the circuit connectors for ease of weighing before

References p. 61

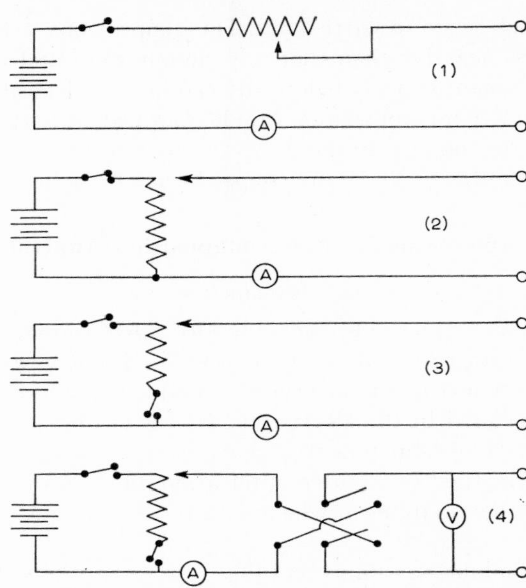

Fig. II, 3. Circuits for quantitative electrodeposition, (1) for simple determinations, (2) for separations by potential control, (3) for both purposes, (4) for both purposes with change of polarity and central-zero voltmeter.

and after deposition. The distance between the electrodes should be small, and vigorous stirring should be possible; sometimes this has been arranged by using a rotating anode. Suitable designs have been described by Sand[24, 25] and Fischer[26]. Other very similar designs have been produced by apparatus manufacturers.

The reasons for choosing gauze are that electrodes in this form have a greater surface area than they would have if made from the same amount of platinum in any other form, and that more free circulation of electrolyte is possible with consequent reduction of electrolysis time and production of even deposits. A recent improvement has been introduced by replacing the gauze by expanded platinum sheet, which has a greater strength than woven wire mesh and has similar area per unit weight[27]. The frame material of both these mesh electrode types is not pure platinum, but a harder alloy of platinum with other metals having similar resistance to chemical attack.

The practice of drying electrodes over a Bunsen flame after washing deposits is not recommended, since some deposited metals readily alloy with platinum and do this much more rapidly at elevated temperatures. To minimise this possibility, the deposition of mercury, tin, lead, cadmium or zinc is often made upon electrodes previously coated with copper or silver. Several useful types of electrode system now available are shown in Fig. II,4.

Electrodes should not be deformed, and should therefore be handled by means of tongs or forceps by the stem only. If deformation occurs accidentally, the use of cylindrical wooden formers to re-shape the wire gauze is recommended. Platinum electrodes are best cleaned by heating in strong nitric acid and, when in regular use, should be kept immersed in dilute nitric acid.

Other metals that have been used for cathodes are tantalum, silver (especially for copper determinations), nickel (for zinc determinations), copper (for copper determinations) and amalgamated brass (for zinc determinations). The use of metals other than platinum for anodes is less satisfactory, although some use has been made of passive iron and of graphite.

Liquid cathodes, especially mercury and, in some cases, Wood's metal have specialised uses, later to be described.

Stripping solutions are used to remove deposits and many variations of method are possible. It is important to avoid attack of the electrode. Hot, slightly diluted nitric acid is used to remove copper, bismuth and lead; hot, slightly diluted hydrochloric acid is used for zinc and tin, and strong ammoniacal trichloroacetate for copper (especially from silver electrodes). Other stripping solutions are described in connection with individual determination techniques.

(d) Arrangement of Equipment

The complete apparatus for quantitative electrodeposition by manual control can be assembled from parts easily available or readily made in chemical laboratories, or an outfit may be purchased complete.

Whichever method is chosen, the finally assembled equipment should have facilities for continuous control of potential between electrodes up to at least six volts, a current-carrying capacity of up to ten amperes, suitable electrode clamps to hold the electrodes steady and make good electrical connection, and also to allow rapid disconnection of the cathode. Facility for rapid stirring and heating of the container is also essential. When separations by potential control are employed, facilities for holding an auxiliary electrode close to the cathode and for measuring the potential difference between it and the cathode are also necessary. As electrolysis vessels, tall-form beakers without lips are recommended, and for most electrode systems, a nominal capacity of 120 ml is satisfactory. When in use, the beaker should be covered with a split watch glass to catch spray.

References p. 61

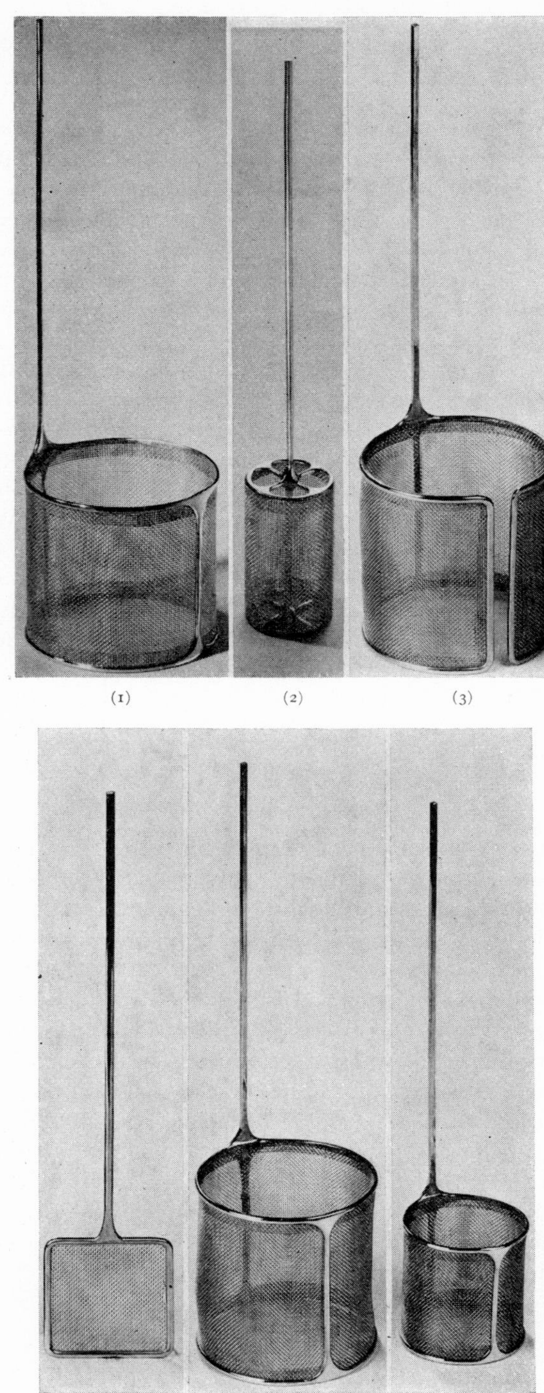

Fig. II, 4. Platinum electrodes used in quantitative electrodeposition. (1), (3), (5) cathodes, (2), (4) rotating anodes, (6) fixed anode.

5. Technique for Single Determinations without Potential Control

Sequence of Operations

Adjust the stirrer speed to about 300 r.p.m., and the level of the beaker support so that the electrode system is just above the bottom of the beaker. Fit the electrodes after weighing the cathode, raise the beaker with the analytical sample in solution into position, fill up with water *just* to cover the electrodes, place a split watch glass over the beaker, and commence stirring. Raise the temperature to the required value, switch on the current and adjust it to the correct value and allow electrolysis to proceed for the required time (to complete exhaustion of the electrolyte). Wash down the sides of the beaker, the exposed parts of the electrodes and the watch glass and continue electrolysis for a further period (about $\frac{1}{4}$ of the original time). Leaving the current and stirrer on, lower the beaker while washing the electrodes, switch off, disconnect the cathode, rinse it rapidly in two successive containers of acetone and dry it in an oven or in a stream of hot air. Cool and reweigh the cathode.

Typical Method

To determine copper in the absence of metals with similar deposition potentials, for example copper in brass, dissolve up to 1 g of the sample in the minimum amount of nitric acid, dilute, add 1 ml concentrated nitric acid and 1 ml concentrated sulphuric acid, dilute to about 100 ml and electrolyse at room temperature at about 5 A. When the solution is colourless, wash down, disconnect and weigh as described above. The added nitric acid prevents reduction of cupric ions to cuprous at the cathode and so shortens the deposition time.

6. Technique for Separations by Manual Potential Control

In these, the potential of the cathode with reference to the solution is controlled by using an auxiliary electrode. Sand's original method[24] is used with only small changes in modern practice. Essential connections are shown in Fig. II,5. The battery B is allowed to discharge through the resistance *R*,

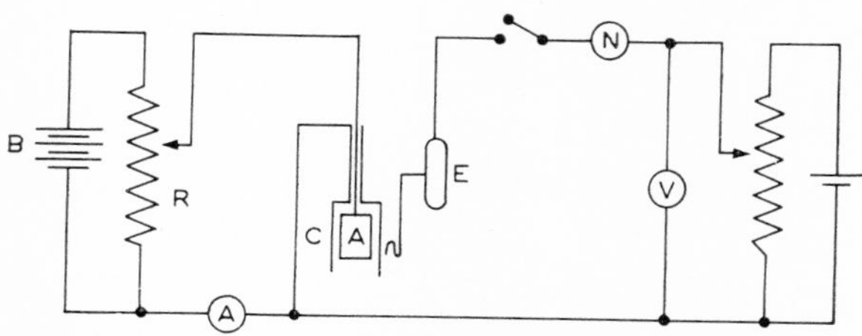

Fig. II, 5. Sand's circuit for electrodeposition with potential control.

References p. 61

from which any potential from zero to the full value of the battery can be tapped off and a current passed through the electrolysis cell A,C, in which a stirring anode is assumed to be used. The potential of the cathode C, is measured with reference to the auxiliary electrode E, by means of a suitable instrument. In the diagram, a potentiometer is shown with a null-point instrument N (Sand in 1907 employed a capillary electrometer), but more modern equipment would employ a valve voltmeter or a *very high resistance* moving-coil voltmeter.

(a) Auxiliary Electrodes

A suitable vessel for the auxiliary electrode on an adjustable stand is shown in Fig. II,6, but other designs may be used if they are of sufficiently low resistance in comparison with the measuring circuit. They should also be designed to minimise diffusion of test electrolyte into the side arm and to flush out the small amount that does diffuse at the end of a determination. It is important that the contact end of the auxiliary electrode should be placed outside the lines of electrolytic current flow so that the measurement of e.m.f. does not include a voltage drop due to the current passing.

Suitable half cells with their potentials are shown in Table II,5.

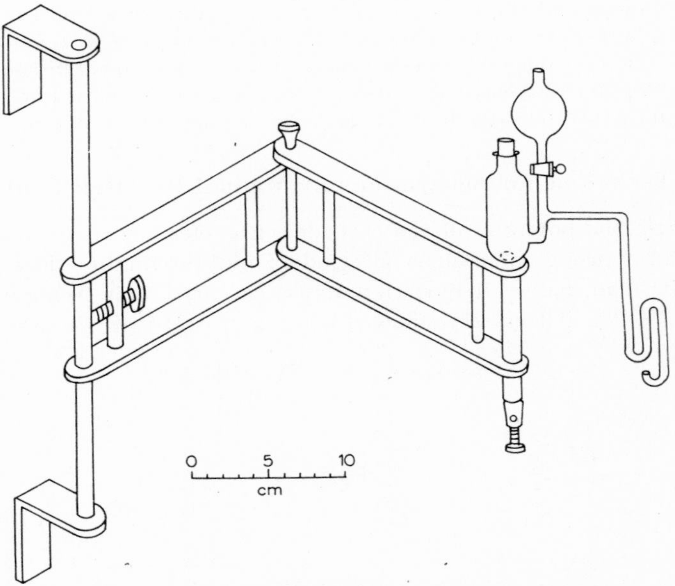

Fig. II, 6. Auxiliary electrode and adjustable stand.

TABLE II, 5

REFERENCE HALF-CELLS AND THEIR POTENTIALS AT 20°

Half-Cell	*Potential* ($E_H = 0$) (*volts*)
Hg | Hg_2Cl_2 | KCl (saturated)	+0.25
Hg | Hg_2Cl_2 | KCl (solid) | $NaNO_3$ (50 g/100 ml)	+0.26
Hg | Hg_2Cl_2 | KCl (solid) | Na_2SO_4 (20 g hydrate/100 ml)	+0.25
Hg | Hg_2SO_4 | H_2SO_4 (2*N*)	+0.68
Ag | AgCl | KCl (0.1*N*)	−0.13

(*b*) *Potential Measurement*

Measurement of the potential of a working electrode with respect to an auxiliary electrode was formerly effected by balancing it *via* a null-point instrument such as a galvanometer against an adjustable known potential. More recently, a high-resistance direct current voltmeter or a valve voltmeter has been employed for this purpose, and the latter, because of its very high impedance, can be very accurate, even with a moderately high resistance-measuring circuit. The valve voltmeter also has the advantage that it can be the controller for an automatic system for adjustment of electrode potential.

(*c*) *Sequence of Operations*

Adjust the stirrer speed to about 300 r.p.m. and the level of the beaker support so that the electrode system is just above the bottom of the beaker. Fit the electrodes after weighing the cathode. Add the necessary reagents such as acid or depolariser to the analytical sample, raise the beaker with the solution into position, fill up with water just to cover the electrodes, insert the auxiliary electrode, cover with a split watch glass and commence stirring. Heat to the required temperature, switch on the current and increase the potential difference between the electrodes until the cathode assumes the required deposition potential. Maintain this potential until the current falls to a low value, wash down the exposed parts of the electrodes, the watch glass and upper parts of the beaker, flush out the auxiliary electrode and then raise the potential by 0.05 v. Allow the current to fall again, increase the potential by 0.05 v again and allow the current to fall for a third time. Lower the beaker with continuous washing, disconnect the cathode, rinse it in two successive containers of acetone and dry it in an oven or in a hot air stream. Reweigh the electrode with the deposit. Further metals may be deposited in turn from the residual electrolyte.

A modified technique was used by Brown[28] who employed a wire of the metal being deposited, or a platinum wire coated with the metal (prepared by putting the platinum wire in parallel with the cathode), as an auxiliary electrode. The wire is then used with a high-resistance voltmeter between

References p. 61

it and the cathode to measure the potential of the latter. The potential difference between the wire and the cathode is usually small, and approximately represents the difference between the reversible potential of the metal in the solution and the deposition potential. This arbitrary value has to be determined by a trial experiment, but once found it is a satisfactory control. The wire and the cathode are weighed together before and after the deposition, so that any metal dissolved off the wire and re-deposited on the cathode may be accounted for. Fig. II,7 shows a suitable clamp for the Brown wire electrode. The stem of this is of insulating material and of a size suitable for it to fit into the holder for the auxiliary electrode shown in Fig. II,6.

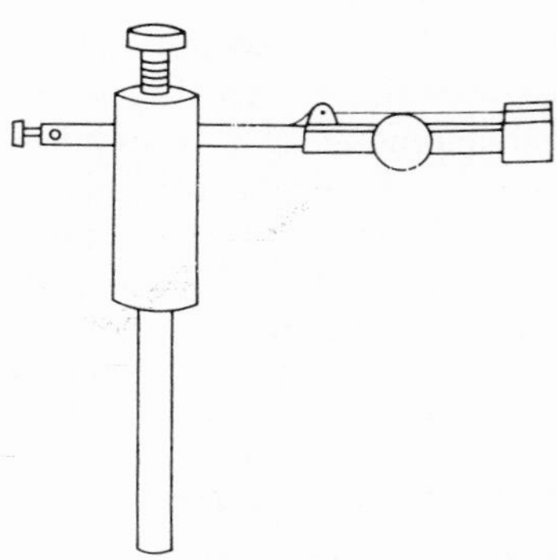

Fig. II, 7. Support for Brown's wire auxiliary electrode.

7. Technique for Separations by Automatic Potential Control

Automatic potential control is the natural labour-saving consequence of the manual methods described above and, in recent years, represents the major advance in electrodeposition analysis. The principle of this method is that a constant potential is maintained between the working electrode and a reference or auxiliary electrode immersed in the solution. Exhaustion of the solution, or other causes tending to alter the potential of the working electrode, are made to change the applied voltage *via* an electronic circuit or a servo-mechanism driving a control resistance, so that a constant potential is maintained between the working electrode and the solution.

(a) Electronic Potential Controllers

For electrolysis currents not exceeding a fraction of an ampere, the "potentiostat" of Hickling[29] is satisfactory, or a more elaborate device described by Greenhough, Williams and Taylor can be employed[30]. Such devices are very satisfactory for microchemical electrodepositions.

An entirely electronic apparatus for potential-controlled separations has been described[31]; this keeps the potential difference between the cathode and a reference electrode within 5 mV of a chosen value and provides up to 8 A to the electrolysis cell. It is driven from the alternating current mains and employs semiconductor devices throughout. It is compact and self contained (Fig. II,8).

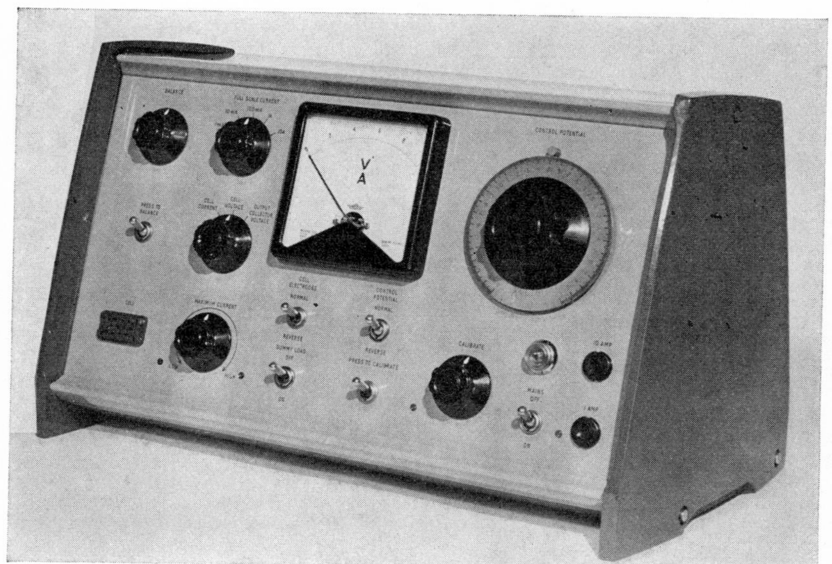

Fig. II, 8. Wadsworth's electrodepositor.

(b) Electro-mechanical Controllers

For the larger currents employed in macrochemical processes (0.1 to 1.0 gram), a servo-mechanical control is satisfactory. These devices are almost as numerous as the electrochemical laboratories employing deposition techniques. Generally, changes in potential between the cathode and the auxiliary electrode are made to control a rheostat or autotransformer supplying the electrolysis current.

The principle of a circuit devised by Lingane and Jones[32] is shown in Fig. II,9. The working potential of the cathode with reference to the auxiliary electrode AE, is selected by the potentiometer P, and voltmeter V_1, and the electrolysis is initially controlled by a manually operated "Variac" MV. If the voltage differs at any time by an amount greater than 10 mV

References p. 61

from the preset value, a double pole galvanometer relay R_1 operates, and starts a mains-driven motor M, through an electronic relay R_2, to raise or lower the potential at the electrode to within less than this difference. The motor is geared mechanically to drive another "Variac" AV, in the main supply to the electrolytic cell, consisting of a transformer T, full-wave rectifier R, and a filter F.

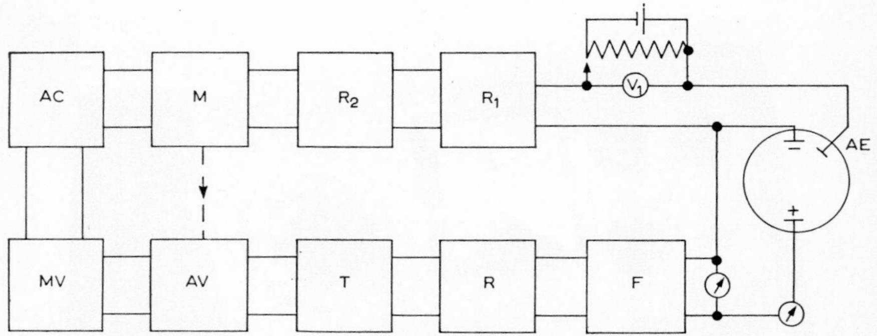

Fig. II, 9. Principle of control circuit of Lingane and Jones.

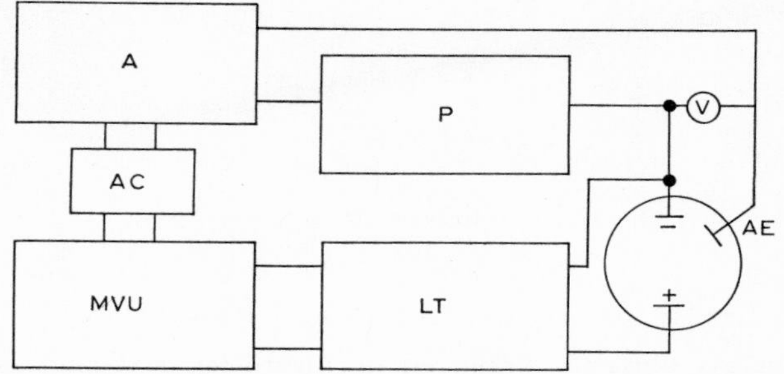

Fig. II, 10. Principle of control circuit of Caldwell, Diehl and Parker.

A very satisfactory instrument for automatic control of potential in electrodeposition analysis was described by Caldwell, Diehl and Parker[33, 34] and a similar arrangement was used by Palmer and Vogel[35], who employed British components and thus made an instrument that could be readily reproduced in Britain and other non-dollar areas. The principle of this circuit is shown in Fig. II,10.

This instrument is operated entirely from the electricity supply mains, which provide the electrolysis direct current by means of a motor-driven

"Variac" unit MVU, and a low voltage supply unit LT, comprising a full-wave metal rectifier and smoothing circuits. The voltage between the cathode and the auxiliary electrode AE is measured by means of a valve voltmeter V, and can be counterpoised by means of a potentiometer P. Differences between the potentiometer voltage and that of the electrode are amplified at A and used to control the "Variac" and thus to keep a constant potential at the cathode. Control is sensitive to 1 mv.

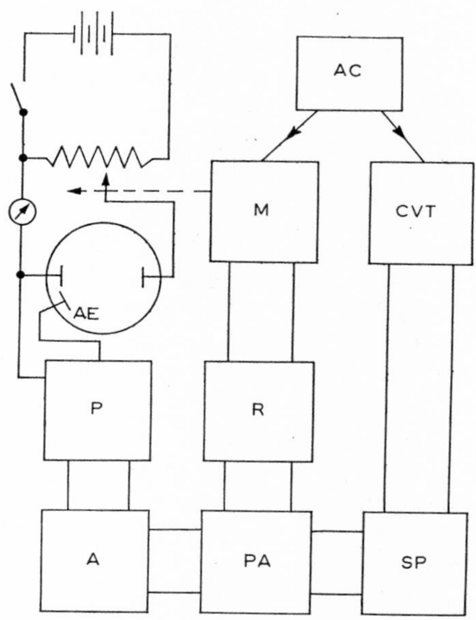

Fig. II, 11. Principle of control circuit of Milner and Whittem.

Another useful circuit was described by Milner and Whittem[36] and the essential parts are shown in Fig. II,11. This instrument has the advantage that it will correct for changes in voltage in both directions and is sensitive to ± 5 mv. It has the small disadvantage that it does not run entirely from the electricity supply mains and so needs a number of batteries.

The e.m.f. between the cathode and the auxiliary electrode AE is balanced by a potentiometer P at the commencement of the determination, and the changes in potential that ensue are amplified in two stages, first in an amplifier A, which then controls a paraphase amplifier PA. The output of this passes through the relays R, controlling the electric motor M. Finally, the motor is mechanically coupled to the slider on a rheostat in the elec-

References p. 61

trolysis circuit, thus compensating for the change in cathode e.m.f. Stability is ensured by use of a constant voltage transformer CVT, and a stabilised power supply SP, for the amplifiers.

8. Conditions of Deposition for Quantitative Analysis of Metals

In deciding upon a method of analysis by electrodeposition, it is necessary to know the nature of the elements present other than the one to be determined, and then to decide if separation of the required element is possible by simple deposition or by some other procedure such as the use of complexing reagents. Potential control is often an advantage, even when it is not necessary as a separation technique, for it frequently assists the production of a deposit that is compact and adheres well.

In this section, detailed instructions are given for the determination of the commoner metals, sometimes by several methods. The methods include most techniques and can be extended readily to metal determinations not described in detail. Thus, where descriptions are given without potential control, they can be adapted for this technique by a trial experiment with an auxiliary electrode. For metals other than those described in detail, or for modifying the methods given for use in very different media or in the presence of other constituents, it is valuable to perform a trial electrolysis by raising the potential difference between the electrodes until deposition commences, and then to plot the current flowing against the potential between the cathode and the auxiliary electrode. Such deposition curves prepared under experimental conditions for analytical work are valuable guides to subsequent separation procedures.

Several other important practical points are briefly discussed below. For potential-controlled separations, rotating stirring anodes are advantageous, since they are usually more efficient than are separate stirrers; closer potential control is possible with better stirring, and so more precise separations are possible and more even deposits are formed. With such electrode systems, washing and disconnection are also easier. The difficulties of making good electrical connections to a rotating electrode have been emphasised, but bearings with independent brush connections for the current supply are a feature of some good commercial instruments[37].

It is also important to recall that metals with closely similar deposition potentials are to be regarded as one analytical group and cannot be separated from each other by simple potential control.

Since the saturated calomel electrode with a suitable junction electrolyte is the most commonly used auxiliary electrode, all potentials given are of the working electrode with reference to this standard. Junction potentials between different electrolytes are small and can be ignored for analytical separations by potential control.

Deposition from hydrochloric acid solution is of more general use than from any other medium, but it must be remembered that the use of an efficient anodic depolariser (to prevent anodic attack by chlorine) is essential with all chloride solutions.

In the details of methods that follow, quantities are given for 100 ml of analytical solution, and the elements are discussed in order of the E_0 values of simple ions as given in Table II,1 (p. 9), and thus in the order of the arbitrary group classification. The quantities of reagents added are generally suitable for any volume between 100 and 120 ml. The common reagents employed are:

1. Acetic acid, 5*N*
2. Ammonia solution, sp.gr. 0.880.
3. Ammonium chloride.
4. Ammonium oxalate.
5. Chromium trioxide, 10% aqueous solution.
6. Hydrazine, 50% aqueous solution.
7. Hydrochloric acid, concentrated, sp. gr. 1.16.
8. Hydrofluoric acid, 40% aqueous solution.
9. Hydroxylammonium chloride.
10. Nitric acid, concentrated, sp. gr. 1.42.
11. Oxalic acid.
12. Perchloric acid, 20% aqueous solution.
13. Potassium chlorate.
14. Potassium cyanide, 10% aqueous solution.
15. Sodium acetate.
16. Sodium hydroxide solution, 5*N*.
17. Sodium peroxide.
18. Sodium sulphate.
19. Sodium sulphite.
20. Sodium tartrate.
21. Sulphuric acid, concentrated, sp. gr. 1.84.
22. Tartaric acid.
23. Zinc chloride.

(a) *Precious Metals Group*

(1) *Gold* can be deposited from acid solutions and also from cyanide solution. The use of this latter medium provides a means of separating it from a number of other metals.

Add 5 ml hydrochloric acid and 1 to 2 g hydroxylammonium chloride. Electrolyse at 25°, + 0.230 V, and 2 to 3 A for about 20 minutes. Note that the auxiliary potential difference is much more negative than would be expected if gold were present as simple ions; it is probably present as complex anions. The presence of these complex ions lowers the cathode potential so much that it is difficult to separate gold from metals of the B group.

Neutralise the gold solution with sodium hydroxide solution and add 1 to 2 g potassium cyanide, electrolyse at 25°, and at 2 to 3 A, for about 20 minutes. Gold deposits are readily removed from platinum by a solution of potassium cyanide and hydrogen peroxide.

References p. 61

(2) *Platinum* can be deposited from acid solutions. In the presence of cyanides it is not readily deposited, and gold can therefore be separated from platinum in cyanide solution.

Add 5 ml hydrochloric acid and 1 to 2 g hydroxylammonium chloride. Electrolyse at 25°, + 0.20 V and 2 to 3 A for about 20 minutes.

(3) *Mercury* can be deposited from nitric or hydrochloric acid solutions or from ammoniacal solutions. If both valency states are present, the electrode potential due to each is the same when the ratio Hg_2^{2+} to Hg^{2+} is approximately 100. In halide solutions, the mercuric ions are present to a much less extent, and so a greater potential is required to effect complete deposition than is necessary in nitrate solution. Mercury tends to alloy with platinum, and so it is often deposited upon silver-plated platinum cathodes. Care must be exercised to avoid losses of mercury by volatilisation or by mechanical means.

Add 1 to 2 ml nitric acid and electrolyse at 50° and 2 to 5 A.

Add 5 ml hydrochloric acid, 1 to 2 g hydroxylammonium chloride and electrolyse at 90 to 100°, and 2 A for about 20 minutes.

Add 10 ml nitric acid, 20 ml ammonia solution and electrolyse at 90 to 100°, + 0.10 V and 2 A for about 10 minutes.

(4) *Silver* can be deposited from acidified nitrate or acetate solution, neutral cyanide solution or ammoniacal solution. From nitrate solutions at lower temperatures, there is a tendency for higher oxide deposition on the anode. Also from nitrate solutions crystalline, loosely adherent deposits can be formed. Satisfactory quantitative determinations can be made at higher temperatures and by co-deposition with mercury.

Add 1 to 2 ml nitric acid and a known amount of a standard solution of mercury as nitrate. Electrolyse at 90 to 100° and 2 A for about 20 minutes.

Add 5 ml nitric acid and 25 g ammonium acetate. Electrolyse at 90 to 100° and 2 A with potential control.

Add 3 ml nitric acid, ammonium hydroxide to neutralise and 2 to 3 g potassium cyanide. Electrolyse at 90 to 100° and 2 A for about 20 minutes.

(b) Copper Group

(1) *Copper* can be deposited from acidic sulphate, nitrate and chloride solutions and also from alkaline cyanide and ammoniacal solutions. Because of the considerable difference in electrode potentials of the two valency states, cuprous ions are always present in low concentration, but in sufficient quantity to react with hydroxyl ions at high pH to deposit cuprous oxide

on the cathode and thus to give a high analytical result. Copious liberation of hydrogen, especially towards the end of a determination, is possible and may lead to the production of a powdery non-adherent deposit.

Add 1 ml sulphuric acid and 1 ml nitric acid and electrolyse at 25° and 5 A for about 15 minutes. A silver cathode can be used.

Add 5 ml hydrochloric acid and 2 g hydroxylammonium chloride, electrolyse at 70°, −0.4 V, and 5 A for about 15 minutes.

Add ammonia solution till the precipitate first formed redissolves and electrolyse at 25° and 2 A for about 20 minutes.

(2) *Bismuth* may be deposited from nitric acid, nitrate, chloride, sulphate, oxalate or tartrate solutions. Spongy deposits are often formed, in part due to heavy evolution of hydrogen when no potential control is exercised.

Add 3 ml nitric acid and 0.2 g hydrazinium sulphate. Electrolyse at 80 to 85° 1.5 A and −0.02 to −0.17 V for about 25 minutes.

Add 10 ml hydrochloric acid, 5 g oxalic acid and 0.5 g hydrazinium chloride. Electrolyse at 80 to 85° and −0.15 V for about 20 minutes.

Add 20 ml sulphuric acid and 1 g hydrazinium sulphate. Electrolyse at 90 to 100°, −0.05 to −0.15 V and 1.2 A for 20–30 minutes.

Add 2.5 ml nitric acid and 12 g sodium tartrate. Electrolyse at 90 to 100° for about 20 minutes.

(3) *Antimony* is deposited most readily from hot hydrochloric acid solutions; it should be in the tervalent state.

Add 10 ml hydrochloric acid and 1 g hydroxylammonium chloride. Electrolyse at 70° and −0.40 V for 20 minutes. Deposits are removed by heating in a mixture of dilute nitric acid and tartaric acid.

(4) *Arsenic* cannot be deposited quantitatively from solution, except with copper in quantity about five times as great as the arsenic. The arsenic must be in the tervalent condition and the amount is obtained by subtracting the quantity of copper added from the weight of the deposit.

To the solution containing about 0.05 g arsenic add 15 ml hydrochloric acid, 1 g hydrazinium sulphate and a known quantity of a standard solution of cupric chloride or sulphate containing about 0.3 g copper. Electrolyse at 50°, −0.4 V and 4 A for about 20 minutes.

(c) Lead Group

(1) *Lead* is deposited cathodically as metal or anodically as dioxide (see p. 44). Coarsely crystalline deposits of metal are produced from nitrate solu-

References p. 61

tions, but smooth deposits from chloride or tartrate solutions in the presence of anodic depolarisers. Since lead readily alloys with platinum, the electrode should be plated with copper or silver before use.

Add 20 ml hydrochloric acid and 2 g hydroxylammonium chloride. Electrolyse at 70°, −0.6 V and 2 A for about 30 minutes. Cool before disconnection and dry and weigh quickly to avoid high results due to oxidation of the deposit.

Add 1 ml nitric acid, 20 g tartaric acid and make alkaline with ammonia solution. Electrolyse at 95 to 100° and 2 A for about 20 minutes.

(2) *Tin* may be deposited from hydrochloric acid solutions of both stannous and stannic salts, although it is likely that the quadrivalent tin is cathodically reduced to the bivalent state before deposition. It is also considered that very stable quadrivalent complexes may be formed when the simple salts are left in solution for prolonged periods[38].

Add 10 ml hydrochloric acid, 1 g ammonium chloride and 4 g hydroxylammonium chloride. Electrolyse at 70°, −0.6 V and 2 A for about 30 minutes. Before disconnecting, the solution should be made just alkaline with ammonia solution to minimise the risk of re-dissolving the deposit[39, 40]. As tin alloys readily with platinum, it is preferable to deposit tin upon copper-plated platinum.

Add 5 ml hydrochloric acid, 10 g ammonium oxalate and 2 g hydroxylammonium chloride. Electrolyse at 70° and 3 A for about 20 minutes.

(d) Zinc Group

The metals of this group are readily attacked by dilute acids and, although they may deposit well, or even quantitatively, from acid solutions, they redissolve very easily and this process is strongly assisted by electrolytic action between the platinum electrode and the deposit during washing and disconnecting. Generally, it is preferable to deposit metals of this group from solutions of controlled pH. Amalgamated brass cathodes are also useful[41]. For deposition of some metals in this group, ammoniacal solutions are very suitable. A selection of methods follows.

(1) *Nickel and Cobalt* can be quantitatively deposited from ammoniacal or oxalate solutions.

Add 25 ml ammonia solution, heat to 75° and electrolyse at 5 A for about 20 minutes.

Add 15 g ammonium oxalate, heat to 95 to 100° and electrolyse at 5 A for about 30 minutes.

(2) *Cadmium* can be deposited quantitatively from alkaline, neutral or

acid solutions; in the last case, deposition is only possible because of the high hydrogen overvoltage of the element.

Add 1 ml hydrochloric acid and 2 g hydroxylammonium chloride. Electrolyse at 25° and 1 A for about 40 minutes.

In the absence of halides, add sodium hydroxide solution to precipitate the metal hydroxide, just re-dissolve with acetic acid, add 5 g sodium acetate and adjust the pH to 4.5. Electrolyse at 80°, with potential control (−0.62 V) and 2 A for about 30 minutes.

Add sodium hydroxide solution to precipitate the metal hydroxide and just re-dissolve the precipitate with potassium cyanide solution. Electrolyse at 50 to 60°, −0.65 V and 2 A for about 20 minutes.

(3) *Zinc* can be deposited from solution in which the pH is controlled. Quantitative deposition on solid electrodes is impossible from strongly acid solutions, although a mercury cathode or an amalgamated brass cathode can be used. It is recommended that platinum gauze electrodes should be coated with copper or silver before use, because zinc readily alloys with platinum.

Add sodium hydroxide solution until the precipitate first formed is re-dissolved, add a few drops of hydrazine solution and then acetic acid until the precipitate first formed re-dissolves. Electrolyse at 25° and 2 A for about 30 minutes.

Add 5 ml acetic acid, a few drops of bromophenol blue and then sodium hydroxide solution until the solution is violet. Electrolyse at 25° and 3 A for about 30 minutes, keeping the colour violet by addition of more indicator and sodium hydroxide as required.

Add ammonia solution until a precipitate starts to form, then 1 g potassium cyanide and 5 ml more of ammonia. Electrolyse at 25° and 3 A for about 30 minutes.

(e) Mercury Cathode Group

Whereas a mercury cathode can be used for deposition of metals from all the foregoing groups, it has special advantages in analysis of metals with more negative deposition potentials than those of Group D. Outstanding advantages are, *(i)* an amalgam is formed when the metal deposits; thus there is no difficulty of the kind experienced with solid electrodes in obtaining adherent deposits; *(ii)* for the same reason, the deposit is protected from oxidation; *(iii)* the very high hydrogen overvoltage of mercury allows the deposition of metals that could not be plated on to platinum electrodes. The mercury cathode also has well established disadvantages namely, *(i)* very small droplets of mercury are easily lost during washing and drying of the amalgam; *(ii)* loss by volatilisation is also possible, but in most analyt-

References p. 61

ical conditions is negligible[42]; *(iii)* the used mercury in the form of amalgam has to be purified, but good methods are available.

Very many types of containing vessel have been described for the mercury cathode, and their design has been influenced by the type of analytical process they were required to fulfil. Gravimetric determinations require small quantities of mercury and the facilities for complete separation of the cathode amalgam in a dry condition for weighing; for this purpose, the simplest vessels are often the best, and the use of a tall-form beaker with a platinum contact wire sealed through the bottom is very satisfactory. Frequently,

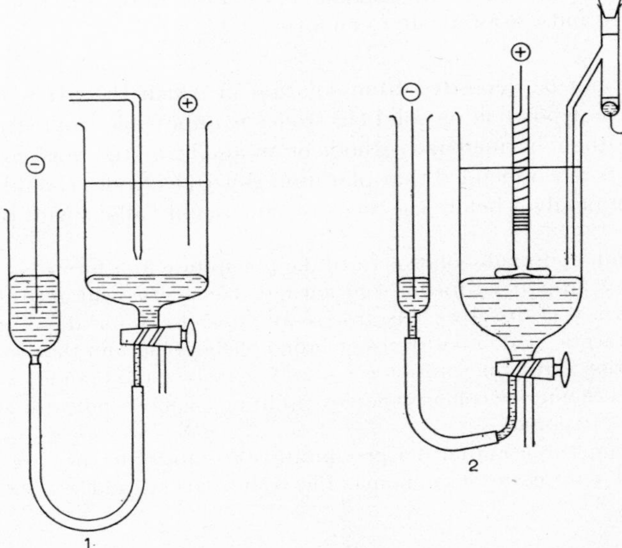

Fig. II, 12. Mercury cathode electrolysis apparatus (1) without, and (2) with cathode potential control.

the vessel and the mercury are weighed together before and after electrolysis, but there is always risk of entraining electrolyte or washing water to give excessive weight, or of losing amalgam droplets to give deficient quantities. Thus, special care is necessary in using this kind of apparatus.

More recently, the gravimetric determination of metals as amalgams has become less common, but the use of the mercury cathode as a means of separation prior to other methods of determination has been developed extensively. A very considerable advantage of this method of removing constituents is that the introduction of impurities by added chemical reagents is avoided. The most generally useful type of cell is that of Melaven[44] and various modifications of it[45, 46]; two of these are shown in Fig. II,12.

Another type of apparatus that is especially useful is the immersion or "unitized" electrode system, which can be employed in a tall-form beaker for the electrolytic removal of a large number of elements before the determination of others[47, 48]. Two such electrode systems are shown in Fig. II,13.

The use of cathode potential control by means of an auxiliary electrode is a very important development of the mercury cathode separation technique. This method has been developed especially by Lingane and may be combined with a polarographic study, from which preliminary information upon the chemical composition and the electrical constants of the electrode system

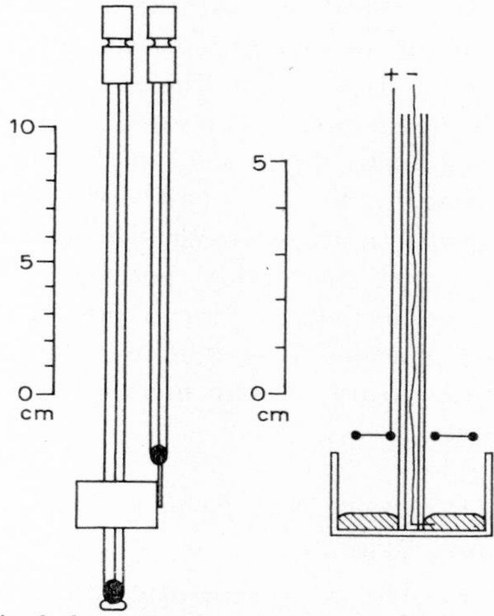

Fig. II, 13. Unitized electrode systems for mercury cathode electrolysis.

can be obtained. It is then possible to adjust the electrolytic potential so that all unwanted elements are removed, thus leaving the electrolyte available for other chemical or polarographic determinations.

Quantitative deposition of a large number of metallic elements from aqueous solutions on liquid mercury has been reported: bismuth, cadmium, chromium, cobalt, copper, gallium, germanium, gold, indium, iridium, iron, mercury, molybdenum, nickel, palladium, platinum, polonium, rhenium, ruthenium, silver, technetium, thallium, tin and zinc have all been determined this way, and a number of other elements have been separated but not quantitatively determined.

Most mercury cathode methods are employed at the present time for separations of more noble metals from others prior to their determination by non-electrolytic means. However, some electrolytic gravimetric deter-

References p. 61

minations are employed and methods employing the Paweck electrode are especially useful. A selection of mercury cathode techniques is given below, illustrating various types of method. Reference may be made to three important reviews [50, 51, 52].

(1) *Use of the Paweck Electrode*

The amalgamated brass gauze electrode first used by Paweck[41] is effectively a mercury cathode, and is especially useful for gravimetric depositions of less noble metals *e.g.* those of Group D. The Paweck electrodes are readily made from brass gauze and stout brass wire. The simplest way to amalgamate the brass electrode is to dip it into saturated mercuric chloride solution to which a few per cent of nitric acid has been added. A grey deposit is produced which becomes bright when the electrode is heated in dilute hydrochloric acid. Most of the metals deposited on such a cathode may be removed by heating with hydrochloric acid in the same way, and the electrode may then be used again. Nickel, cobalt, cadmium and zinc may all be determined quantitatively on the Paweck electrode using the conditions given on pp. 32 and 33 with, preferably, potential control to prevent the deposition of still less noble metals. Special precautions are necessary in drying the electrode to avoid loss of weight by volatilisation of mercury; a current of air previously saturated with mercury vapour has been recommended for this purpose.

(2) *Mercury Cathode Depositions from Small Volumes*

Two useful electrolytic cells suitable for smaller volumes (3–50 ml) are illustrated in Fig. II, 14. The first[53] employs a tap funnel with a two-way stopcock and a platinum wire anode, and is especially useful for the removal of iron from small steel samples before the determination of other metals. The second[54] employs a silver foil cathode amalgamated by electrolysing mercurous nitrate solution. The anode is a platinum wire, as is also the lead to the cathode, which is spot-welded to the silver. For this apparatus, electrolysis currents of up to 1 A have been used to separate copper and iron from aluminium, uranium and zirconium.

(3) *Aluminium, Magnesium and Beryllium in Nickel Alloys*

This method[47] was developed for the analysis of the special alloys of nickel used in the manufacture of electrodes for radio valves and similar electronic devices.

Dissolve 2 g of the sample in a mixture of 8 ml water and 8 ml nitric acid, add 10 ml sulphuric acid, evaporate till fumes of sulphuric acid appear, heat

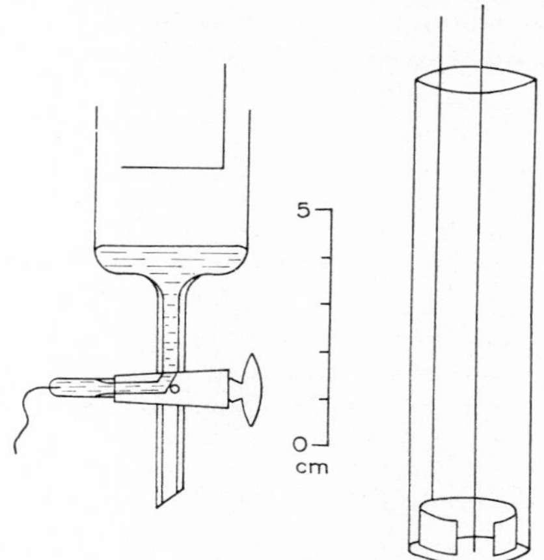

Fig. II, 14. Small-scale electrodeposition systems for mercury cathodes.

for a further 10 minutes, cool and dilute with 50 ml water, boil till all sulphates have dissolved, filter off silica and wash. Add sodium hydroxide solution (20%) until a permanent turbidity is produced, redissolve this with sulphuric acid and add an excess of 1 ml. Dilute to 150 ml, heat to boiling and electrolyse with the apparatus shown in Fig. II, 13, at 8 A for 75 minutes. The nickel and any more noble metals are removed. The electrolyte contains aluminium, which can be precipitated and weighed as Al_2O_3 after ignition at 1200° for 30 minutes.

The same separation method may be used for nickel-magnesium and nickel-beryllium alloys.

(4) *Aluminium, Calcium and Sodium*[49]

Dissolve the sample in hydrochloric or a mixture of hydrochloric and nitric acids, add 5 ml sulphuric acid and concentrate to small bulk and evolution of sulphuric anhydride. Dilute and electrolyse for about 30 minutes at 5 A. Determine aluminium, calcium and sodium in residual liquid by non-electrolytic means.

(5) *Aluminium Alloys containing up to 10% Aluminium*

Dissolve the specimen in 20 ml of a mixture of equal volumes of hydrochloric acid, nitric acid and water, add 10 ml perchloric acid and evaporate to remove all the less volatile acids, dilute to 100 ml, filter if necessary from any silica and electrolyse with mercury cathode at 15 A for 30 to 40 minutes. Cd, Co, Cu, Fe, Mn, Mo, Ni, Pb, Sb, Zn, Cr are removed; Ti, V, W, Ta, Mg, Si, Nb, Th, U, Zn, Ce and Al remain. Extraction with chloroform of the 8-hydroxyquinaldinates at pH 9.2 followed by further extraction of the 8-hydroxyquinolinates at

References p. 61

pH 9.2 leaves only aluminium, and this may be determined spectrophotometrically as the quinolate. The method[55] has been successfully applied to iron, steels, aluminium bronzes and brasses.

(6) *Steel and Iron*

Dissolve the sample in 12 ml sulphuric acid (1 : 9) for every 1 g taken, filter off undissolved carbides, wash, neutralise with sodium hydroxide solution and add excess acid of 4 ml for each 100 ml solution. Electrolyse with the mercury cathode for about 2 hours at 15 A (test for iron should be negative) and then for a further 15 minutes. Aluminium, titanium, vanadium, zirconium, calcium and magnesium may be determined in the residual liquid by non-electrolytic means[56].

(7) *Separation of Lanthanons from one another*

Some useful analytical techniques have been reported by Onstott[57, 58] using mercury cathodes, although the lanthanons are very reactive elements. Thus, europium can be separated quantitatively from samarium on a lithium amalgam electrode by electrolysing an aqueous solution of the two in lithium citrate. The cathode potential remains effectively constant because of the constant composition of the lithium citrate.

9. Separation of Mixtures of Cations

A very large number of methods have been described for the determination of metals in the presence of others either singly or serially, and the application of this technique to alloys and industrial mixtures has become important. In this section, a representative selection of methods has been chosen and they are arranged in the order of Table II,1 (p. 9). In cases where several electrolytic metal determinations are described in succession, the most noble metal analysed determines the order of description. Although it is generally true that metals in the same analytical group (see section 3, p. 14) cannot be separated from each other by potential control, there are a number of variables other than electrode potential that may make separations possible. Additionally, it may be useful to remove several metals quantitatively together and to weigh the mixture and then, after redissolving, to use other methods (*e.g.* colorimetry) to determine constituents in this mixture. The use of a mercury cathode (section 8*e*) has become increasingly useful in removing a number of more noble metals from light alloys containing aluminium and magnesium.

In all determinations described in this section, the total electrolysis volume is 100 ml unless some other volume is explicitly prescribed. Auxiliary electrode potentials quoted are those with reference to the saturated calomel electrode.

(a) Precious Metals Group

Mercury and Silver are not separable from each other by simple potential control and, if present together, will be co-deposited. They are, however, readily separable from Groups B, C, D, *etc.* (Table II, 3, p. 14).

(b) Copper Group

(1) *Copper and Antimony*

These metals cannot be separated by simple deposition and, from hydrochloric acid solutions, they are deposited together (p. 30). They may, therefore, be determined together as a mixed deposit and so separated from other metals, for example, tin and lead in bearing-metal alloys. They are separated from the mixed deposit by converting the antimony to a complex antimonic fluoride[59].

Dissolve the deposit in a mixture of 5 ml nitric acid and 5 ml hydrofluoric acid. Quickly dilute to about 80 ml, add aqueous chromic acid solution until the solution is yellow, dilute and electrolyse at 1 A and 25° without potential control. Weigh the copper deposited and determine the antimony by difference.

(2) *Alloys of Copper, Antimony, Lead and Tin (white metals, bearing metals)*[60]

Weigh from 0.2 to 0.4 g of the alloy as drillings, dissolve it by warming with a mixture of 10 ml hydrochloric acid, 10 ml water and 1 g ammonium chloride, adding drop by drop a saturated solution of potassium chlorate. Boil off excess of chlorine, add 5 ml hydrochloric acid, 1 g hydrazinium chloride, dilute to 100 ml and electrolyse at 75° and −0.4 v. Copper and antimony are deposited together. After 15 minutes flush out the auxiliary electrode and continue electrolysis for a further 5 or 10 minutes. Dry and weigh the copper and antimony. Dissolve the deposit in 5 ml nitric acid, 5 ml hydrofluoric acid and 10 ml water. Boil off oxides of nitrogen, dilute to 100 ml and add chromic acid till the solution is yellow. Electrolyse at 25° and −0.4 v, for 20 minutes and weigh the copper. The weight of antimony is obtained by difference.

To the residual liquid add 1 g hydrazinium chloride and deposit tin and lead together by electrolysing at 25° and −0.7 v for 20 minutes. Weigh both metals together. Dissolve the deposit in a mixture of 15 ml nitric acid, 5 ml hydrofluoric acid and 15 ml water, boil off oxides of nitrogen and deposit the lead as dioxide using the outer electrode as anode as described on p. 44. For alloys low in lead, insufficient lead will be present for quantitative deposition of lead as dioxide from this solution. In such cases, add 20 ml of a standardised 0.5% solution of lead as nitrate in order to get complete deposition, and determine the lead in the sample by difference.

References p. 61

(3) *Alloys of Copper, Lead, Tin, Nickel and Zinc with small amounts of Iron, Aluminium and Manganese (nickel bronzes)* [61]

Weigh out from 0.5 to 1 g of the alloy, heat it with 10 ml hydrochloric acid and 1 g ammonium chloride, adding a saturated solution of potassium chlorate drop by drop and boiling after each addition. Evaporate to small bulk, add 5 ml hydrochloric acid and 1 g hydrazinium chloride and dilute to 100 ml. Deposit copper at 50° and −0.4 V. Weigh the copper. Deposit tin and lead together at 25° and −0.7 V in about 20 minutes. Determine the lead as dioxide as described on p. 44. Evaporate the residual liquid with 3 ml sulphuric acid till fumes appear, dilute to 50 ml, make alkaline with ammonia solution and filter off iron and aluminium oxides. Determine these metals colorimetrically. Add 20 ml ammonia solution and 2 g sodium sulphite and deposit nickel at 70° and 1 to 1.1 V for 20 minutes. Cool to 25°, add 5 ml ammonia solution and deposit zinc at 3 A. Heat the residual liquid to boiling and precipitate manganese by passing hydrogen sulphide for 5 minutes. Filter off the sulphide, dissolve it in nitric acid and determine manganese colorimetrically as permanganate using bismuthate.

(4) *Alloys of Copper, Lead, Tin and Zinc (brasses and bronzes)*

Dissolve 0.2 to 0.4 g of the alloy in 10 ml hydrochloric acid, 2 ml nitric acid, 10 ml water and 1 g ammonium chloride. Add 5 ml hydrochloric acid and 1 g hydrazinium chloride, and dilute to 150 ml. Electrolyse at 50° and −0.4 V with initial current 3 to 4 A, falling to 0.1 A. All the copper with any arsenic present is deposited. Weigh the deposit. Dissolve it in 5 ml sulphuric acid, 5 ml nitric acid and 10 ml water. Boil off oxides of nitrogen, dilute to 150 ml and electrolyse for 20 minutes at 25° and −0.4 V. Only copper is deposited. Tin and lead are determined as described for bearing metals. The solution remaining after these metals have been removed is oxidised by boiling with excess of bromine until the colour has gone. Add ammonia till the solution is just alkaline to phenolphthalein and filter off the ferric hydroxide. Determine the traces of iron colorimetrically as thiocyanate after re-solution. Add 10 ml ammonia solution and electrolyse at 25° and 3 A for 20 minutes. Weigh the zinc deposit.

(5) *Alloys of Copper, Nickel and Zinc (nickel silver)* [62]

Dissolve 1 g of the alloy in 4 ml nitric acid, 4 ml sulphuric acid and 10 ml water, evaporate to remove nitric acid (white fumes); dilute and deposit copper, with or without potential control. Neutralise residual liquid with ammonia solution, add 2 g sodium sulphite and 20 ml additional ammonia solution, and electrolyse at 70° and −0.25 V. After weighing the nickel, electrolyse again at 25° without potential control at 2 A for about 30 minutes.

(6) *Copper and Iron*[63]

Although the deposition potentials of copper and iron are so different, their electrolytic separation is difficult, especially when only traces of copper are present. The reason is that ferrous ions are oxidised at the anode and the reduction potential of these ions at the cathode is less negative than the deposition potential of the copper. The use of a diaphragm anode and hydrazine in the catholyte makes copper deposition possible.

To a solution containing up to 10 g iron add 2 ml sulphuric acid and 0.5 ml hydrazine solution. Electrolyse using diaphragm anode containing 10% sodium sulphate at 25° and −0.4 V. After 15 minutes flush out the anolyte with 5 ml of fresh sodium sulphate solution. Electrolyse for a further 15 minutes and separate, wash and weigh the deposit.

(7) *Bismuth and Lead in Nitrate Solution*[64]

Deposit the bismuth in the presence of hydrazine (p. 31), add 50% sodium hydroxide solution until the precipitate first formed redissolves, and destroy the residual hydrazine by adding 1 g sodium peroxide in small portions with boiling to expel evolved gases. Neutralise with nitric acid and add an excess of 10 ml. Electrolyse at 90 to 95° and 5 A with the outer electrode as anode. Weigh as PbO_2 using appropriate conversion factor (0.8635 for 0.3 g., p. 44).

(8) *Bismuth, Lead and Tin in Chloride Solution*

Determine the bismuth by the method given on p. 31, then the tin and lead together as on p. 39. Finally determine the lead in the mixed tin-lead deposit as dioxide after re-solution in hydrofluoric acid and nitric acid, p. 44.

(9) *Fusible Alloys containing Bismuth, Lead and Tin*[64]

Dissolve about 0.1 g of drillings by adding 1 to 2 ml nitric acid and, when the reaction has abated, 10 ml hydrochloric acid and boiling. After adding a further 5 ml hydrochloric acid, dilute to 100 ml, add 5 g oxalic acid and 0.5 g hydrazinium chloride. Electrolyse at 80 to 85°, −0.15 V (initial) to −0.30 V (final) for about 20 minutes.

From the residual liquid, deposit tin and lead together at 80 to 90° and 2 A for about 30 minutes.

Dissolve the deposit in 5 ml hydrofluoric acid, 15 ml nitric acid and 15 ml water, boil to remove oxides of nitrogen, dilute and deposit the lead as dioxide on the anode at 90 to 95° and 5 A for about 20 minutes.

References p. 61

(10) *Antimony and Lead*

These are not easily separated in solutions of simple ions, although from hydrochloric acid solutions with less than 0.2 g lead present this is possible (see p. 39).

Deposit the lead as dioxide at 90 to 95° and 2 A for about 20 minutes from a solution containing 5 ml hydrofluoric acid and 20 ml nitric acid, to which enough chromic acid has been added to give a yellow solution.

(c) *Lead Group*

(1) *Lead and Tin*

These metals are deposited together from chloride solutions and may be weighed together (p. 39).

Dissolve the mixed deposit in 15 ml nitric acid, 5 ml hydrofluoric acid and 15 ml water, boil to remove oxides of nitrogen, dilute and deposit the lead as dioxide at 90 to 95° and 5 A for about 20 minutes.

(2) *Solders*

Solders are generally alloys of tin and lead; sometimes antimony, bismuth and cadmium also are present. The analytical methods are those given for these elements.

(d) *Zinc Group*

(1) *Nickel and Cobalt*

These cannot be separated by simple potential control but the latter can be deposited as Co_2O_3 on the anode if a diaphragm is used to keep the cathode liquid separate[66]. The revolving diaphragm electrode devised by Sand[67] is made the cathode.

Dissolve the sample in hydrochloric acid with the aid of small additions of chlorate. Evaporate almost to dryness on a water bath, add 2 g sodium sulphate, 10 g ammonium chloride and 20 ml ammonia solution; dilute to 150 ml and electrolyse at 70° and −0.7 V for 20 minutes. Weigh the deposit of cobalt and nickel. Dissolve in a mixture of 5 ml nitric acid and 10 ml water, evaporate to dryness, dissolve in water and deposit the cobalt as oxide using the diaphragm electrode as described on p. 45. Obtain the nickel by difference.

(2) *Nickel and Zinc*

To the solution containing both metals in quantity up to 0.2 g and in the form of sulphates, add 5 g tartaric acid, 5 g ammonium chloride, 2 g sodium sulphate and 20 ml ammonia solution. Electrolyse at 70° and −1.0 V for 5 minutes, then raise the potential to −1.1 V and continue electrolysis for 15 minutes more. Separate, wash, dry and weigh the deposit of nickel.

Cool to 25°, add 5 ml ammonia solution and electrolyse at 3 A for 25 to 30 minutes. Separate, wash, dry and weigh the zinc deposit.

(3) *Cadmium and Zinc*

These are separated by adjustment of acidity and by potential control.

To a solution containing about 0.3 g of each metal, add 10 ml hydrochloric acid and 1 g hydroxylammonium chloride; add thymol blue and adjust by adding sodium hydroxide solution to about pH 1.5. Electrolyse at 25°, −0.8 V and current of about 1 A.

To the residual liquid add sodium hydroxide to pH 6 (yellow to methyl red), 1 g potassium cyanide and 10 ml ammonia solution, electrolyse at 3 A for about 30 minutes.

(4) *Aluminium Alloys*

The following procedure is applicable to aluminium alloys containing copper, zinc, nickel, and silicon with traces of iron, tin, lead, manganese and magnesium[68].

Dissolve from 0.5 to 1.0 g of the alloy in 10 ml hydrochloric acid and 20 ml of water, warming if necessary; the trace of black residue is then taken up by boiling and adding a few drops of saturated potassium chlorate solution. Cool, add 15 ml sulphuric acid and evaporate to fuming; cool, dissolve in 30 ml water, add 10 ml hydrochloric acid and boil. Filter off the silica, wash with hot dilute acid, ignite and weigh. If the alloy contains more than 5% of lead, some lead sulphate may be filtered out with the silica. Determine the metals copper, tin and lead in the filtrate as described for nickel bronzes (p. 40). Add 5 to 10 g tartaric acid to the residual liquid, make just alkaline with ammonia solution and add excess of 20 ml. Add 2 g sodium sulphate and electrolyse at 70° and −1.0 to −1.1 V for 20 minutes. Weigh the nickel. Cool, add 5 ml more ammonia solution and deposit the zinc at 20° and 3 A for 20 to 30 minutes. Determine iron and manganese in the residual solution by saturating with hydrogen sulphide and filtering off the sulphides. Aluminium may be determined with 8-hydroxyquinoline.

10. Anodic Deposition of Oxides

Several metals can be deposited as higher oxides on an inert anode. Lead is especially noteworthy in this connection, since the process can be made quantitative without special precautions. The anodic reactions proposed for this process are,

References p. 61

$$Pb^{2+} \rightarrow Pb^{4+},\ Pb^{4+} + 4OH^- \rightarrow Pb(OH)_4 \rightarrow PbO_2 + 2H_2O$$ [69].

The deposit is always somewhat hydrated and the last traces of water cannot be removed readily; however, if deposition is effected at near boiling point, only small amounts of water are present, and this proportion is almost constant for a given set of conditions. An empirical factor for conversion of the weight of deposit to the weight of lead present is therefore always employed.

Other metals deposited anodically as oxides probably involve similar mechanisms, and deposits are often not precisely of the composition expected for the anhydrous oxide. Cobalt, thallium and plutonium have been determined satisfactorily as higher oxides, empirical conversion factors being used when necessary.

(a) Determination of Lead

Make the outer electrode the anode in the electrolytic cell. Add 10 ml of nitric acid to the solution of lead as nitrate, dilute to 100 ml and electrolyse at 90 to 95° and 5 A for 10 or 15 minutes. Remove the anode and dry and weigh as for metal deposits. The theoretical factor for calculation of Pb from PbO_2 is 0.866. Experimentally determined factors making allowance for the small amounts of moisture are as follows:

Amount of Lead	*Factor*
up to 0.1 g	0.8660
0.1 to 0.4 g	0.8635
0.4 to 0.5 g	0.8605

The deposit is readily removed by boiling the electrode in a mixture of oxalic acid and dilute nitric acid.

(b) Determination of Cobalt

In order to deposit cobalt quantitatively upon an anode, it is necessary to use a cathode separated from the anolyte by means of a diaphragm, and to avoid strongly acidic solutions. The revolving diaphragm electrode[70] is illustrated in Fig. II,15. It consists of a silver rod sealed into a glass tube with solder; the lower end of the glass tube is secured by means of four glass beads to an outer tube with a thistle-funnel top, and the gauze electrode fits over the lower end of this outer tube. A parchment diffusion thimble with two longitudinal slits cut in its upper part is fitted over the electrode and secured round the funnel neck by means of a rubber band. By pouring fresh electrolyte into the funnel, the original contents are flushed out together with any electrolyte that may have diffused in from the outer compartment; thus the outer electrolyte may be quantitatively exhausted with respect to the ions being determined.

To the solution containing less than 0.04 g cobalt add enough sodium hydroxide solution to neutralise, then 1 ml acetic acid and 10 g sodium acetate. Dilute to 100 ml and electrolyse at 90 to 95° with initial current 0.5 A increased to 1 A after 10 minutes. Use as cathode the diaphragm electrode filled with 10% nitric acid. Flush out the catholyte with 5 ml of 10% nitric acid and continue electrol-

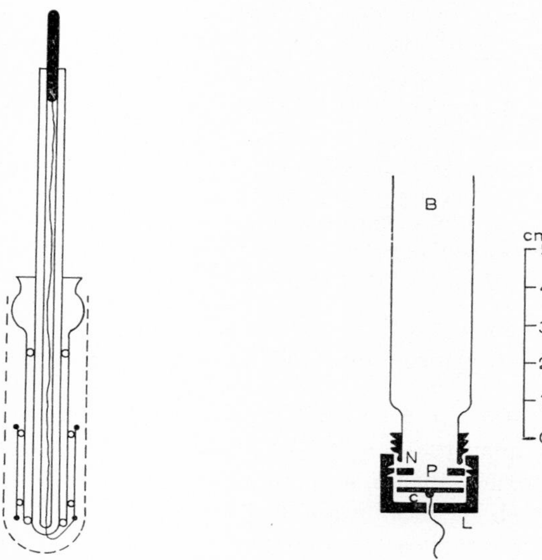

Fig. II, 15. Sand's revolving diaphragm anode.

Fig. II, 16. Apparatus for preparation of radioactive electrodeposits.

ysis for another 20 minutes. Weigh the deposit after separating the electrode and drying in the usual way. Deposits may contain occluded salts and, to ensure accuracy, they may be redissolved and deposited again. Warm the deposit with 20 ml of 20% sulphuric acid to which a few drops of saturated sulphur dioxide solution have been added. Boil to remove excess, neutralise with ammonia solution and add an additional 20 ml, dilute and electrolyse with ordinary electrodes for 20 minutes at 3 A. The theoretical conversion factor of 0.710 may be used[71].

(c) *Determination of Thallium*

To the solution of thallous sulphate or nitrate add 10 ml nitric acid and 0.1 g cupric nitrate. Make just alkaline with ammonia solution, add an extra 5 ml, and electrolyse at 2 A for about $\frac{3}{4}$ hour. Weigh the deposit of Tl_2O_3 after drying at 160° and employ the factor 0.895. Traces of copper, which is added to prevent

References p. 61

cathodic deposition of thallium, may be found in the deposit. Manganese, lead and bismuth interfere in this determination[72].

(d) Determination of Plutonium

The deposition of plutonium as PuO_2 from solution has been employed for quantities from 0.5 μg up to 2 mg on platinum disc anodes of a suitable size to give reproducible quantitative determinations by counting techniques[73].

The electrolytic cell is shown in Fig. II,16. It can be made from a cylindrical screw-capped bottle B with the bottom cut off. The platinum disc P rests on a thicker metal disc C and both fit the lid L, which is bored to take the connecting wire. A neoprene washer N makes the liquid-tight joint with the glass. A spiral stirring cathode is used.

Oxidise the plutonium in 2*N* sulphuric or hydrochloric acid by passing ozone until it is all in the plutonyl state, neutralise the solution with potassium hydroxide solution and add an excess until the whole is 2*N*. Electrolyse at current density of 40 mA per square cm. The yellow deposit of PuO_2 is washed, dried and then counted.

11. Microchemical Electrolytic Analysis

This is, in general, a transfer from ordinary scale (decigrams) down to about one-hundredth quantities (one to ten milligrams). The electrodes are, however, reduced to only about one-tenth of the area. A considerable number of empirical methods have been described for specific purposes, but the more generally useful are those that allow for potential control and thus for separations by potential adjustment. The methods and general precautions of microchemical techniques are observed. A metal (brass or aluminium) counterpoise is used for the electrode, which is weighed upon the microchemical balance to within a few micrograms. Thus, with a deposit of a few milligrams, an accuracy of weighing of one part in a thousand is possible. The total error is likely to be greater than this, but the method is very satisfactory for analyses in which only a few milligrams are available or desirable.

(a) Apparatus for Potential-controlled Depositions

Potential control by means of an auxiliary electrode with small-scale apparatus introduces considerable difficulties, although a satisfactory silver/silver chloride electrode has been described[74] and is shown in Fig. II,17. An alternative method of cathode potential control is to use an anode much larger than the cathode, and to hold its potential constant by means of a large excess of a suitable anodic depolariser[75]. In these conditions, the anode-cathode potential difference can be used as a control for the cathode process, and separations of the metal groups are possible.

The apparatus shown in Fig. II,18, employs gas stirring and the electrodes are held apart by the gas inlet tube and glass rod spacers. The outer electrode, which is permanently attached to the frame, is normally the anode and the inner electrode the cathode. The success of the method, especially for separa-

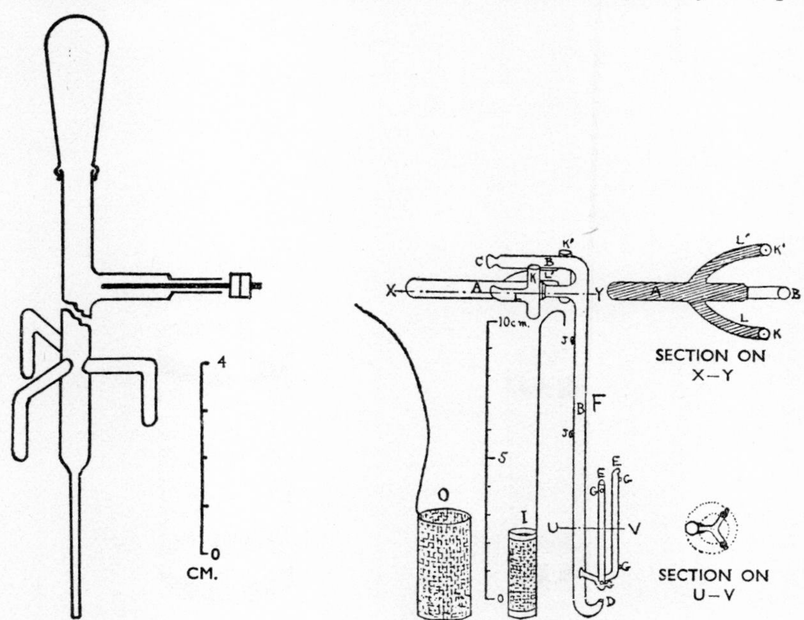

Fig. II, 17. Auxiliary electrode vessel for microchemical electrodepositions.

Fig. II, 18. Apparatus for microchemical electrodepositions.

tions by potential control, depends entirely on efficient stirring, and this in turn depends upon the electrodes being at the bottom of the vessel, which should be of correct size. A fast stream of small stirring gas bubbles can be produced by a partially closed screw clip near the gas inlet.

The method of construction of this apparatus has been fully described[75], but is briefly repeated here. F is the frame on which the electrodes are placed and it is built upon the gas inlet tube B. Stirring gas enters at C and emerges at the capillary jet D. A glass rod A, which carries two arms L,L′ bearing glass cups K,K′ for mercury, serves to support the frame on the stand. The removable electrode I fits easily into the space between a frame consisting

References p. 61

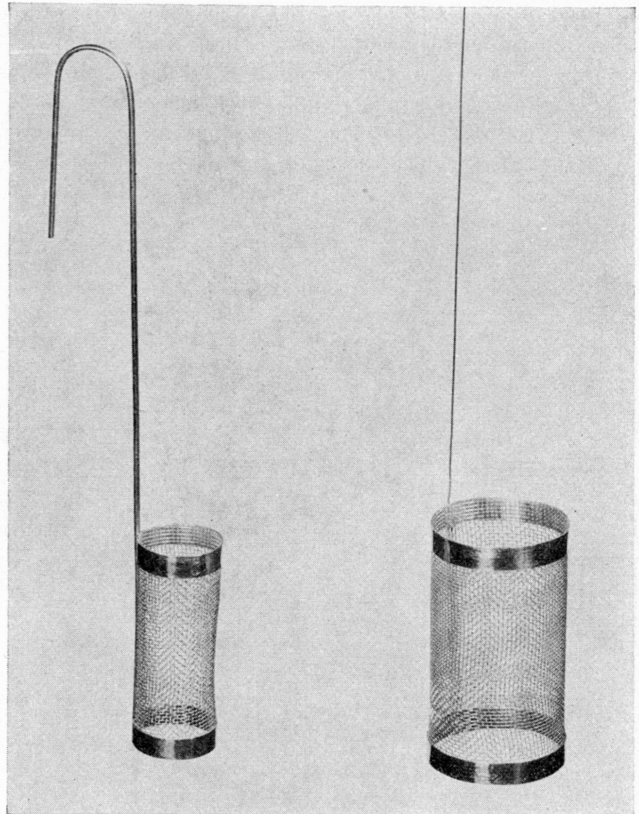

Fig. II, 19. Microchemical platinum electrodes.

of the gas tube B and the retaining rods E, while the fixed electrode O is placed round this frame. The electrodes are shown also in Fig. II,19 and they make contact with the circuit by means of wires dipping into the mercury cups. The electrolysis vessel is a test-tube of internal diameter 2 cm, cut to about 10 cm in length, and holds about 12 ml to the top of the electrodes.

With the electrodes in position, the solution to be analysed is placed in the test-tube, which is raised from below until the frame touches the bottom. The water bath is next raised into position and the gauze-covered ring is swung into position. Water is then added until the electrodes are covered, and the stirring gas (hydrogen, nitrogen or carbon dioxide) is admitted. With the temperature adjusted, the current is switched on and controlled until the correct deposition potential is attained. This value is maintained until the current has fallen to about one-tenth of its original value. After the vessel interior has been washed down with a fine water jet, the current is maintained for a few minutes and then the hot water bath is replaced by a bath of cold water. Disconnection is effected by lowering the test-tube slowly with the voltage still applied and spraying the electrodes thoroughly with a fine water

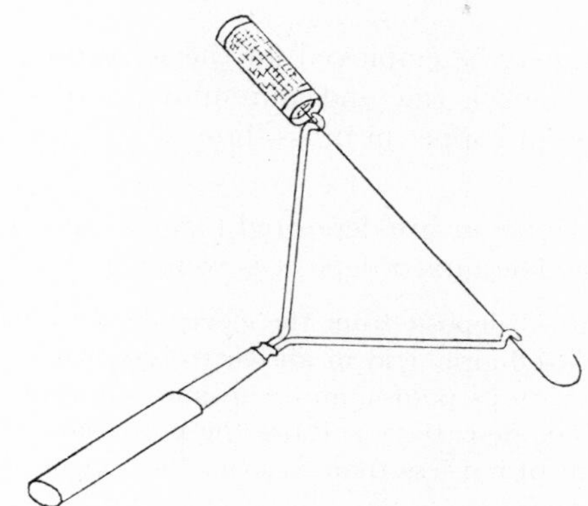

Fig. II, 20. Carrying tool for micro-cathodes.

jet. The inner electrode is then removed with a platinum hook, dipped in turn into two successive baths of acetone, and dried in a stream of warm air. The end that has dipped into the mercury cup is then heated to redness to remove traces of mercury and, after cooling, the electrode is transferred to the microchemical balance, acclimatised for a few minutes and weighed. A convenient tool for carrying the electrode and transferring it to the microbalance is shown in Fig. II,20.

(b) Determinations Using Potential Control

As is necessary with procedures in microchemical analysis, fuller details are given than with full scale methods. In the methods that follow, conditions have been adjusted to suit the apparatus described in the previous section. If the micro auxiliary electrode is employed, deposition potentials as used in macro determinations may be employed, but it should be noted that the decinormal silver chloride electrode has a potential of —0.13 volt on the hydrogen scale.

(1) *Copper* may be determined from acidic sulphate, nitrate or chloride

References p. 61

solutions or from ammoniacal solution. The most satisfactory method is to use potential control.

To the solution containing up to 6 mg of copper add 1 ml hydrochloric acid, 0.1 g hydroxylammonium chloride and dilute to 15 ml. Electrolyse at 60 to 70° at an anode-cathode potential of 1.2 V for 10 minutes, wash down and continue electrolysis for another 5 minutes. Cool, and terminate the electrolysis as usual. If hydrazine is used as depolariser, the potential difference should be 0.7 V.

This method [76] may be employed for the separation of copper in the presence of lead, tin, nickel, zinc and aluminium, and is therefore suitable for the determination of copper in brass, bronze and gunmetal.

(2) *Copper and Antimony* are deposited together under the conditions given for copper above. The mixed deposit is weighed.

Remove the mixed deposit from the electrode with a mixture of 1 ml nitric acid and 1 ml hydrofluoric acid in an electrolysis tube. Dilute immediately to about 15 ml, add 3 drops potassium dichromate solution and electrolyse for 20 minutes at 1.2 V (anode-cathode). Raise the potential difference, if necessary, to maintain a current of not less than 50 mA. The copper deposit is washed, dried and weighed as usual.

(3) *Brasses and Bronzes. (Alloys containing copper, tin, lead, zinc)* [77].

Dissolve the alloy (in the form of fine drillings) in an electrolysis tube in hydrochloric acid with the addition of small quantities of powdered potassium chlorate. Evaporate in a water bath almost to dryness, add 1 ml hydrochloric acid and 5 drops of hydrazine solution and then enough water to cover the electrodes. Heat to 80° in a water bath, stir with nitrogen or hydrogen and raise the potential difference until deposition proceeds satisfactorily (0.8 V). Electrolyse for about 20 minutes for amounts between 5 and 10 mg. Keep the voltage constant until the current has fallen to about one tenth of its initial value, increase the voltage by about 0.05 V and continue electrolysis until the current is 3 mA. Wash and separate the electrodes; dry and weigh the cathode and copper deposit as usual.

Replace the copper-coated cathode in the electrolyte at 40° and electrolyse at 1.3 to 1.4 V for 15 minutes. Cool by surrounding the tube with a water bath, add 1 drop bromophenol blue indicator and hydrazine solution until a blue colour is formed. Wash, separate, dry and weigh the electrode with tin and lead.

Redissolve the whole deposit in 3 ml nitric acid and 7 drops hydrofluoric acid. Heat to boiling and deposit the lead as dioxide, with the inner electrode as anode and a voltage of about 1.0. Replace the vessel with a fresh tube containing 2 ml nitric acid, 5 drops hydrofluoric acid and enough water to cover the electrode. Dissolve off the deposit of lead dioxide by reversing the current and once more deposit the dioxide by reversing again. Separate, wash, dry and weigh the pure dioxide deposit as described below.

To the solution from which copper, lead and tin have been removed add sodium hydroxide solution until a precipitate is formed. Just redissolve this in acetic acid, the final pH being about 4.6. Heat to 40° and electrolyse at 1.25 V for about 25 minutes. Wash, separate and dry the electrodes and weigh the cathode and zinc deposit.

(4) *White Metals* containing copper, antimony, tin, lead and zinc are analysed in the same manner as described for brasses and bronzes but, as the proportions of the various metals are so different, the conditions are changed somewhat[78].

Dissolve the alloy in the form of fine drillings (5 to 20 mg) in hydrochloric acid in an electrolysis tube, adding small amounts of powdered potassium chlorate. Evaporate almost to dryness in a water-bath, add 1.5 ml hydrochloric acid and 4 drops hydrazine solution. Electrolyse at 80°, and 0.75 V for about 20 minutes with nitrogen or hydrogen stirring. The current should then be about one-tenth of its original value. Raise the voltage by about 0.05 V and continue electrolysis until the current falls again. Weigh the mixed copper and antimony deposit.

Replace the cathode and deposit and, with no applied voltage, surround the electrodes with a tube containing 7 drops nitric acid, 2 drops hydrofluoric acid and enough water to cover the electrodes. Heat the solution to near boiling, and add potassium dichromate to give a definite yellow colour when all the deposit has gone from the cathode. Add enough hydrazine solution to destroy the colour and to give an excess of three drops. Electrolyse for copper at 0.7 V for ten minutes.

Tin and lead are determined exactly as described for bronzes (p. 50).

(5) *Bismuth* may be determined from nitric acid solution[79].

Add 1 ml nitric acid, 2 drops 50% hydrazine solution, and dilute to 12 ml. Heat to 60 to 70° in the water bath and electrolyse at an anode to cathode potential of 0.80 V. The current falls from about 80 to about 10 mA after 10 minutes. Wash the tube down with a fine water jet and increase the potential difference to 0.90 V. After a further 3 minutes cool and disconnect in the standard manner. Deposits are removed from the electrode by means of nitric acid.

(6) *Lead* may be determined as dioxide anodically deposited from nitric acid solution[79].

To the lead solution containing up to 6 milligrams, add 2 ml nitric acid and dilute to 12 ml. Electrolyse with the weighed inner electrode as the anode, at just below boiling temperature, and at 1.0 V and 200 mA. The potential is not critical and may be increased to give the correct current density. After 7 minutes

References p. 61

wash down and continue electrolysis for another 3 minutes. Terminate the electrolysis by rapidly replacing the electrolysis tube with another containing water. The electrolytically deposited lead dioxide is always slightly hydrated and the experimentally determined factor 0.860 is used to calculate the amount of lead.

(7) *Separation of Bismuth and Lead* may be achieved by serial electrolysis.

To a solution containing up to 6 mg of each metal add 1 ml nitric acid and two drops of hydrazine solution. Dilute to 12 ml and determine bismuth as above (Method 5). Wash the deposit by rapidly replacing the electrolyte tube by another containing water. Disconnect, dry and weigh as usual.

Transfer the electrolyte to a 50-ml tall-form beaker using the washing water to complete the transference. Add 50% sodium hydroxide solution until the precipitate redissolves, then add 10–20 mg sodium peroxide and heat until no more oxygen is evolved. Neutralise with nitric acid and add a further 4 ml. Electrolyse at just below boiling temperature at 1.2 V and 200 to 300 mA. The deposit always occludes salts from solution; these are removed as follows: replace the beaker by a tube containing 2 ml nitric acid in 12 ml, reverse the current to remove the deposit and then re-deposit by reversing again. Complete the determination as described for lead in Method 6.

(8) *Solders, Alloys of Tin and Lead.* These metals may be analysed exactly as described under bronzes[77] (See Method 3). Lead alone may be determined as below.

Dissolve the alloy in 3 ml nitric acid mixed with 7 drops of hydrofluoric acid, dilute and electrolyse at the boiling point with the inner electrode as anode at a voltage of 1.0 for 10 minutes. The lead, deposited as dioxide on the anode, is purified and re-deposited by dissolving it into a new electrolyte by current reversal as described under bronzes (Method 3).

(c) Other Techniques

In all the foregoing descriptions, use of the British Standard Apparatus[80] based upon the recommendations of Lindsey and Sand[73] has been advocated. Many other types of apparatus, some not suited to potential controlled separations, have been described and access to these methods may be made through reviews[18, 19, 20, 21, 22, 23, 81].

Ingenious apparatus has been described[82, 83, 84] for determining small amounts of metals in large volumes of solution, and these techniques offer advantages for separating mixtures of metals prior to re-solution and determination by quantitative separation in more manageable volumes for potential control.

A further development is the separation and determination of metals in submicro quantities. Here, the difficulties encountered through incompletely covered electrodes[85, 86] may be reduced by adding known amounts of a metal that co-deposits with the analytical specimen. This method has been especially useful when a radioactive isotope is to be determined, and the only essential in electrodeposition is to have it all deposited with or without carrier metals, because determination is effected by counting techniques.

Typical of such methods is the separation of radio-bismuth (^{212}Bi) and radio-lead (^{212}Pb) from "old" thorium nitrate, in which quantities as small as 10^{-15} g can be determined[87]. In such determinations, it is important that the deposit of carrier and active element should be prepared in a thin uniform film of constant size on one side of a disc electrode, so that the geometry of the specimen and the counter positions may be constant. A cell of the type shown in Fig. II,16 (p. 45) is also satisfactory in this respect.

12. Internal Electrolysis

Although the idea of electrodeposition without an external current source was proposed by Ullgren in 1868, internal electrolysis is a term first used by Sand, and the analytical method was developed by him and his colleagues[88, 89, 90]. Interest has more recently revived in these techniques and additional and improved methods have been proposed[91, 92, 93].

The principle of internal electrolysis is to employ a cell with the anode and cathode compartments kept separate by means of a porous diaphragm. In one compartment, a metal electrode is placed in contact with a solution of its ions and, in the other, a solution of the same metal ions at a lower concentration and with additionally the ions of another metal of more noble (more electropositive) nature. If an electrode of platinum is then inserted in this second compartment and connected to the first electrode, a current will pass and the more noble metal will be deposited upon the inert electrode. The deposition may be made quantitative by stirring the second (cathode) compartment and adjusting the conditions so that the deposit formed is adherent and free from impurities. The formation of poorly conducting films on the anode must be avoided.

The apparatus due to Sand is shown in Fig. II,21, where the cathode, which is weighed before and after electrodeposition, is shown surrounding a centrifugal stirrer; the two anodes, suspended in parchment diffusion shells and immersed in a suitable solution, are supported on either side of the cathode.

A slight hydrostatic pressure is maintained in the anode vessels to hinder the catholyte from diffusing into the membranes.

There is no need to insert a meter between the two electrodes when a well established technique is in use, but a useful circuit to test that the procedure is correct is shown in Fig. II,22. In this circuit, the high-resistance meter serves a double purpose: as a voltmeter with switch 1 open and switch 2 closed, and as an ammeter with switch 1 closed and switch 2 closed.

Internal electrolysis is particularly suitable for the separation of small quantities of a metal from large quantities of a less noble metal. Improvements upon the original method have been the use of collodion-coated anodes (usually of zinc) to dispense with the anode chambers[94].

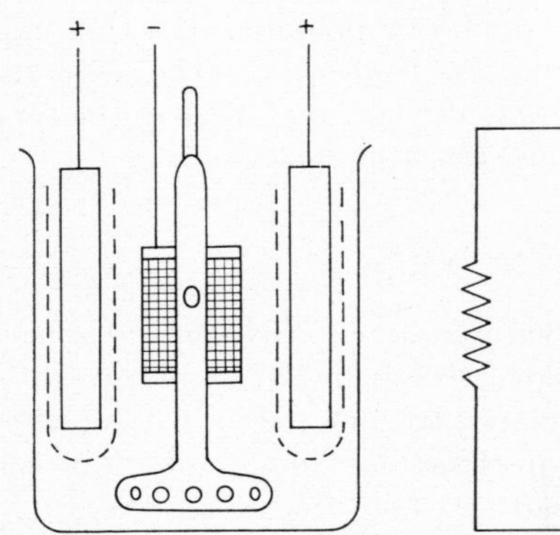

Fig. II, 21. Internal electrolysis system.

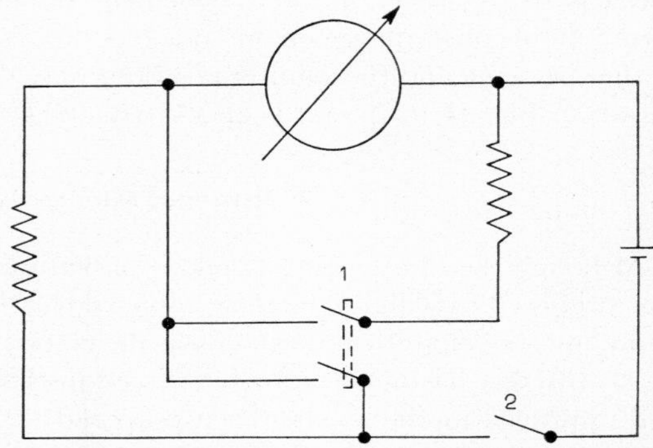

Fig. II, 22. Circuit for internal electrodeposition.

(1) *Silver in the presence of Lead and Traces of Bismuth*[95]

To the solution containing up to 10 g of lead as nitrate and bismuth up to 30 mg as nitrate add 4 ml nitric acid and three drops of hydrazine solution and dilute to 300 ml. Electrolyse at 60° with cast bismuth anodes and an anolyte containing 1 g bismuth and 5 ml nitric acid in 100 ml for about 20 minutes. Disconnect the cathode, wash, dry and weigh. If copper also is present, it will be partially deposited.

(2) *Silver in the Presence of Lead and Traces of Copper and Bismuth*[94]

The method is the same as for silver in the presence of lead and bismuth, except that more nitric acid (9 ml) is added to the catholyte and that the anodes should be of copper immersed in an anolyte containing 1 g of copper as nitrate and a trace of nitric acid in each 100 ml of solution. The method may be employed in the analysis of galena.

The method has been modified for the determination of silver in cupreous pyrites, *i.e.* in the presence of copper, iron, nickel, arsenic and zinc.

(3) *Bismuth and Copper in Lead Bullion*[96]

Dissolve 5 g in 15 ml nitric acid, 1 g tartaric acid and about 50 ml water. Add 2 ml hydrochloric acid, heat to coagulate the silver chloride and filter. Add dilute potassium permanganate solution to the hot filtrate until a faint permanent colour is attained, then add 1 ml hydrazine solution. Dilute to 300 ml and electrolyse at 85° for 15 minutes, using lead anodes and an anolyte containing 2 g lead as nitrate and 5 ml nitric acid in 100 ml of solution. Disconnect the cathode, wash, dry and weigh the mixed deposit of bismuth and copper.

(4) *Copper and Cadmium in Spelter and Zinc Ores*[97]

Dissolve 5 g in 5 ml sulphuric acid with 30 ml water, adding 2 ml nitric acid, if necessary, to complete solution. If lead is present, evaporate to fuming, dilute and filter from lead sulphate. Add 1 ml hydrazine solution and electrolyse for copper at 70° with zinc anodes and an anolyte containing 10 g of zinc chloride and 10 g of ammonium chloride in 100 ml of solution. Disconnect the cathode, wash, dry and weigh.

Adjust the acidity of the catholyte by adding sodium hydroxide solution to give a faint precipitate, and acetic acid until a pH of 5.0 is attained. Electrolyse as for cadmium in zinc (Method 8 below).

(5) *Copper in Iron*[98]

To the solution containing up to 5 g of iron in the ferrous state add 3 ml sulphuric acid and 0.2 g hydrazine sulphate. Dilute to 300 ml and electrolyse at 70° with iron wire anodes and an anolyte containing 5 g iron, 3 ml sulphuric acid and 0.2 g hydrazinium sulphate in 100 ml for 45 minutes. This time is sufficiently long for quantities of copper up to 30 mg. The method has been modified for the determination of copper in steel[99].

(6) *Bismuth in Lead*[100]

To the solution containing up to 1 g lead as nitrate add 2 ml nitric acid, a few drops of hydrazine solution, and dilute to 300 ml. Electrolyse at 85° with lead anodes and an anolyte containing 2 g lead as nitrate and 5 ml nitric acid in 100 ml of solution for 20 minutes. Disconnect the cathode, wash, dry and weigh.

The method can be applied to lead ores such as galena.

(7) *Nickel in Zinc*[101]

To the solution containing up to 5 g zinc as chloride add 30 g ammonium chloride and 2 g sodium sulphate. Dilute to 300 ml and electrolyse at 65° with zinc anodes and an anolyte containing 10 g of zinc chloride, 10 g ammonium chloride and 20 ml ammonia solution per 100 ml for 45 minutes. This time is sufficient for quantities up to 35 mg nickel.

References p. 61

(8) *Cadmium in Zinc*[101]

To the solution containing up to 5 g zinc as chloride add 30 g ammonium chloride, 2 g of sodium acetate, 1 ml 25% acetic acid and 1 g of hydrazinium chloride. Dilute to 300 ml, and electrolyse for 30 minutes at 70° with zinc anodes and an anolyte containing 10 g zinc chloride and 10 g ammonium chloride per 100 ml of solution. Disconnect the cathode, wash, dry and weigh. The time is sufficient for less than 10 mg cadmium, but for greater quantities a longer time is necessary.

13. Electrographic Detection of Metals

This method is an electro-solution technique; the specimen to be tested is made anodic and the electrolyte in its vicinity is tested for cations produced by anodic attack. The electrolyte, hydrochloric acid, ammonium chloride solution or sodium nitrate solution, is poured on to filter paper and the specimen is pressed on to this with a cathode (or counter-electrode) of aluminium on the other side. After the application of a potential difference of a few volts for about 1 minute, the paper is removed and tested by means of suitable reagents for the metal ions. A whole series of specific spot reagents can be applied as drops to identify the metals in the one electrogram. In some cases, it is useful to immerse the whole paper in the reagent solution and thus to "develop" an impression of the specimen. A variation of this technique is to use filter paper soaked in electrolyte mixed with reagent, so that the characteristic reaction of the metal ion is immediately visible. It is thus possible to show variations of composition of metallic articles, the inclusion of other metals, or perforations of plating or other metal finishes. Useful summaries of electrographic methods have been published by Hermance and Wadlow[102, 103].

Very little material is used in the method; the amount may be as little as one microgram.

The method can be performed with a portable instrument powered by dry cells, or with a more elaborate instrument with the facility of varying the pressure between the electrodes and the voltage applied[104]. This latter instrument is powered from the alternating current mains and is fitted with a process timer to ensure standard and repeatable methods.

Methods have been described for chromium, nickel, lead, tin, copper and zinc[105], chromium in steels[106], molybdenum in steels[107] and alloy steels[108].

The use of high pressure to give extremely close contact between the specimen and the test paper is recommended when either detailed prints of imperfections in metal finishes or clear records of inhomogeneity in a metal surface are required. Clearer electrographic records are also produced when gelatine-sized paper is employed in contact with the specimen.

(a) Pore Size Determination in Protective Films by Electrographic Prints[109]

Cut to the same size pieces of blotting paper and of Kodak imbibition paper (double weight) and soak them in 5% potassium nitrate solution for 30 minutes. Remove and squeeze them between sheets of blotting paper at 550 lb/sq. in. Lay the metal specimen on the anode platen, place the smooth side of the imbibition paper on the specimen, then the blotting paper and then the cathode platen. Press to 600 lb., switch on current at 6 v for 20 seconds, switch off release pressure, develop in suitable reagent (*e.g.* 2% potassium ferrocyanide solution for iron) for 30 minutes, wash and dry.

A more elaborate packing of the specimen in the press has been recommended[110], in which a rubber sheet, an aluminium sheet, blotting paper,

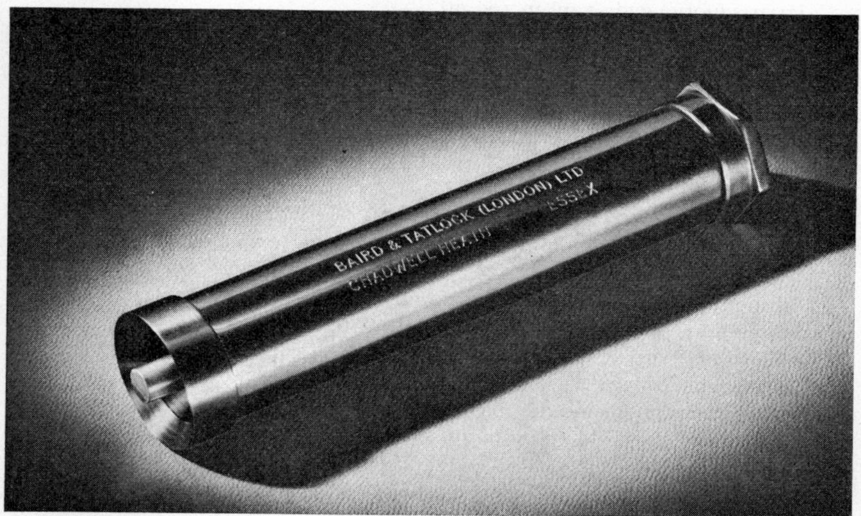

Fig. II, 23. Electrographic spot-test apparatus.

photo paper, the test piece and a second aluminium sheet are placed in sequence in a heavy press. For small specimens, a portable screw-press made from a carpenter's cramp can be used effectively.

(b) Qualitative Detection of Metals in Alloys

The method has been systematised[111] for use with a battery-operated portable apparatus shown in Fig. II,23. A small disc of filter paper soaked in a suitable electrolyte is placed on the object to be tested, the cathode

References p. 61

(the centre spring-loaded electrode) is placed on the paper and the concentric contact ring of stainless steel is pressed down to make contact with the specimen for about 15 seconds. The liberated metal ions are recognised by characteristic reactions. This small and convenient apparatus can be used to explore a considerable area of metal for inhomogeneities.

Details for recognition of various metals are given below. Some interference may be given by other metals (see Vol. IA, p. 382 for spot-test procedures).

(1) *Reagents used in Electrographic Tests*

(a) Electrolytes

Ammonium chloride: 5 g in 100 ml.
Hydrochloric acid: 5 ml in 100 ml.
Phosphoric acid: 25 g in 100 ml.
Sodium dihydrogen phosphate: 5 g in 100 ml.
Sodium fluoride: 5 g in 100 ml.
Sodium nitrate: 10 g in 100 ml.
Sodium potassium tartrate: 10 g in 100 ml.

(b) Developers

Aluminon: 0.1 g ammonium aurine tricarboxylate and 1 g ammonium acetate in 100 ml water.
Ammonium chromate: 0.5 g in 100 ml water.
Ammonium thiocyanate: 1 g in 100 ml water.
α-Benzoin-oxime: 1 g in 100 ml alcohol.
Cacotheline: saturated solution in water.
Cadmium stannous iodide: dissolve 1 g stannous chloride in 2–3 ml hydrochloric acid, add 2 g cadmium iodide and 3 g potassium iodide and dilute to 50 ml.
Cobaltous chloride: 0.05 g in 100 ml 0.5*N* hydrochloric acid.
Dimethylglyoxime: 1 g in 100 ml alcohol.
p-Dimethylaminobenzylidenerhodanine: saturated solution in acetone.
Diphenylcarbazide: 0.02 g and 10 ml acetic acid in 100 ml water.
Formaldoxime hydrochloride: 5 g in 100 ml water.
Mercuric thiocyanate: 8 g mercuric chloride and 9 g ammonium thiocyanate diluted to 10 ml with water.
9-Methyl-2,3,7-trihydroxy-6-fluorone: saturated solution in alcohol or saturated solution in 25% phosphoric acid.
p-Nitrobenzene-azo-resorcinol: 0.025 g in 100 ml *N* sodium hydroxide.
Oxalic acid: saturated solution in water.
Potassium cyanide: 1 g in 100 ml water.
Potassium ferrocyanide: 10 g in 100 ml water.

Potassium iodide: 10 g in 100 ml water.
Potassium thiocyanate: saturated acetone solution.
Thiourea: 10 g in 100 ml water.
Sodium diethyldithiocarbamate: 0.1 g in 100 ml water.
Sodium metabisulphite: 5 g in 100 ml water.
Sodium nitrite: 10 g in 100 ml water.
Sodium sulphide: 10 g in 100 ml water.
Stannous chloride: 1 g in 100 ml 5*M* hydrochloric acid.
Sulphuric acid: 2.5*M*.

(2) *Aluminium*

Use ammonium chloride electrolyte and then aluminon developer; a red colour increasing in intensity indicates aluminium. Iron will interfere; to avoid this difficulty, use ammonium chloride on a strip of paper and add one drop of potassium ferrocyanide before electrolysing. Add aluminon to the side of this drop and water to the centre of the drop; aluminium is indicated by a red colour in the overlap.

(3) *Antimony*

Use phosphoric acid and methyl-trihydroxyfluorone as electrolyte; a bright red colour indicates antimony, if the solution is kept acidic.

(4) *Beryllium*

Use sodium nitrate electrolyte; develop with nitrobenzene-azo-resorcinol, which gives an orange-red with beryllium. Iron, zinc, nickel, copper, cadmium and silver interfere, but their colours are destroyed by adding potassium cyanide solution.

(5) *Bismuth*

Use sodium nitrate electrolyte and develop with thiourea; a yellow colour indicates bismuth. Copper and silver give brown and black colours respectively.

(6) *Cadmium*

Use sodium nitrate electrolyte and develop with sodium sulphide. Metals with dark sulphides interfere.

(7) *Cobalt*

Use sodium fluoride electrolyte and develop with potassium thiocyanate; a brilliant blue-green colour indicates cobalt.

References p. 61

(8) *Chromium*

Use sodium dihydrogen phosphate as electrolyte and add diphenylcarbazide as developer; a violet colour indicates chromium.

(9) *Copper*

Use sodium potassium tartrate electrolyte and develop with potassium ferrocyanide; a brown colour indicates copper. Alternatively, sodium diethyldithiocarbamate, which also produces a brown colour, may be used.

(10) *Iron*

Use ammonium chloride electrolyte and develop with potassium ferrocyanide; a blue colour indicates iron.

(11) *Lead*

Use sodium nitrate as electrolyte and develop with potassium iodide followed by sodium metabisulphite; a yellow colour indicates lead.

Alternatively, develop with sulphuric acid followed by cadmium stannous iodide; an orange colour indicates lead.

(12) *Magnesium*

Use sodium nitrate electrolyte and develop with magneson; a pale blue colour indicates magnesium.

(13) *Manganese*

Use sodium nitrate electrolyte and develop with sodium nitrite followed by oxalic acid; a pink colour indicates manganese.

(14) *Molybdenum*

Use hydrochloric acid electrolyte and develop with stannous chloride followed by ammonium thiocyanate; a red colour indicates molybdenum.

(15) *Nickel*

Use sodium fluoride electrolyte and develop with dimethylglyoxime; a bright red colour indicates nickel.

(16) *Silver*

Use sodium nitrate electrolyte and develop with *p*-dimethylaminobenzylidenes rhodanine; a brownish-violet colour against a yellow background indicates silver- Alternatively, develop with ammonium chromate; a brick-red colour indicate. silver.

(17) *Tin*

Use ammonium chloride electrolyte and develop with cacotheline; a violet colour with a yellowish-brown background indicates tin.

(18) *Zinc*

Use ammonium chloride electrolyte and develop with mercuric thiocyanate, followed by cobaltous chloride; a blue colour indicates zinc.

Acknowledgements

The author wishes to record his indebtedness to the late Dr. H. J. S. Sand, formerly Head of the Department of Physical and Inorganic Chemistry at Sir John Cass Technical Institute, who originated many of the methods of analytical electrodeposition, and who, during years of instruction and collaboration, was a source of inspiration to his many research students. Acknowledgement is made also to some of these and other investigators whose methods are also described in this chapter.

Acknowledgement is also made to the Society for Analytical Chemistry for permission to reproduce diagrams first published in *The Analyst*, to Messrs. Baird and Tatlock (London) Ltd., for permission to reproduce electrographic methods developed in their laboratories and for Figure 23, to Messrs. Johnson Matthey Ltd. for Figure 4, and to Messrs. Southern Analytical Ltd. for Figure 8.

References

1. H. J. S. Sand, *Electrochemistry and Electrochemical Analysis* (Blackie and Son, Ltd., London, 1939). Vol. I.
2. S. Glasstone, *Introduction to Electrochemistry* (D. Van Nostrand Company, Inc., New York, 1942).
3. G. Kortüm and J. O'M. Bockris, *Textbook of Electrochemistry* (Elsevier Publishing Company, Amsterdam, 1951). Vols. I and II.
4. J. O'M. Bockris, *Modern Aspects of Electrochemistry* (Academic Press, New York, 1954).
5. A. Lassieur, *Electroanalyse Rapide* (Les Presses Universitaires de France, Paris, 1927).
6. A. Classen and H. Danneel, *Quantitative Analyse durch Electrolyse* (Springer, Berlin, 7th Edition, 1927).
7. A. Hollard and L. Bertiaux, *Analyse des Métaux par Electrolyse* (Dunod, Paris, 4th Edition, 1930).
8. H. J. S. Sand, *Electrochemistry and Electrochemical Analysis* (Blackie and Son, Ltd., London, 1940). Vol. II.
9. A. Jílek and J. Kofa, *Vázková analysa a elektroanalysa* (Brno-Krávlova Pole, Prague, 1946).

10. W. D. Cooke, *Electroanalytical methods in trace analysis* (Academic Press, New York, 1956). Vol. III.
11. A. Schleicher, *Elektroanalytische Schnellmethoden* (F. Enke, Stuttgart, 3rd Edition, 1947).
12. H. Diehl, *Electrochemical Analysis with Graded Potential Control* (G. Frederick Smith Chemical Co., Columbus, Ohio, 1948).
13. W. Böttger, *Physikalische Methoden der analytischen Chemie* (Akademische Verlagsgesellschaft, Leipzig, 1949).
14. J. J. Lingane, *Electroanalytical Chemistry* (Interscience Publishers, New York and London, 1953).
15. P. Delahay, *New Instrumental Methods in Electrochemistry* (Interscience Publishers, New York and London, 1954).
16. G. W. C. Milner, *Principles and Applications of Polarography and other Electroanalytical Processes* (Longmans, Green and Co. Ltd., London, 1957).
17. P. Delahay, *Instrumental Analysis* (The Macmillan Company, New York, 1957).
18. S. E. Q. Ashley, *Analyt. Chem.*, 1949, **21**, 70.
19. *Idem, ibid.*, 1950, **22**, 1379.
20. *Idem, ibid.*, 1952, **24**, 91.
21. D. D. DeFord, *Analyt. Chem.*, 1954, **26**, 135.
22. *Idem, ibid.*, 1956, **28**, 660.
23. D. D. DeFord and R. C. Bowers, *Analyt. Chem.*, 1958, **30**, 613.
24. H. J. S. Sand, *J. Chem. Soc.*, 1907, **91**, 373.
25. A. J. Lindsey and H. J. S. Sand, *Analyst*, 1934, **59**, 335.
26. A. Fischer, *Z. Elektrochem.*, 1907, **13**, 469.
27. Johnson, Matthey and Co. Ltd., *Platinum Laboratory Apparatus* (London, 1958).
28. J. Brown, *J. Amer. Chem. Soc.*, 1926, **48**, 582.
29. A. Hicklinc, *Trans. Faraday Soc.*, 1942, **38**, 27.
30. M. L. Greenough, W. E. Williams and J. K. Taylor, *Rev. Sci. Instr.*, 1951, **22**, 484.
31. N. J. Wadsworth, R. A. E. Technical Note No. Met. 309. 1960.
32. J. J. Lingane and S. L. Jones, *Analyt. Chem.*, 1950, **22**, 1169.
33. C. W. Caldwell, H. Diehl and R. C. Parker, *Ind. Eng. Chem., Analyt.*, 1944, **16**, 532.
34. H. Diehl, *Electrochemical Analysis with Graded Potential Control* (G. Frederick Smith Chemical Co., Columbus, Ohio, 1948). p. 8.
35. J. F. Palmer and A. I. Vogel, *Analyst*, 1953, **78**, 428.
36. G. W. C. Milner and R. N. Whitten, *Analyst*, 1952, **77**, 11.
37. Baird and Tatlock (London) Ltd., St. Cross Street, Hatton Garden, London, E.C.1.
38. H. J. S. Sand, *Electrochemistry and Electrochemical Analysis* (Blackie and Son, Ltd., London, 1939). Vol. II, p. 70.
39. F. G. Kny-Jones, A. J. Lindsey and A. C. Penney, *Analyst*, 1940, **65**, 498.
40. A. J. Lindsey, *Analyst*, 1950, **75**, 104.
41. H. Paweck, *Z. Elektrochem.*, 1898, **5**, 221.
42. A. H. Porter and F. C. Frary, *Trans. Amer. Electrochem. Soc.*, 1908, **14**, 51.
43. A. E. Pavlish and J. D. Sullivan, *Metals and Alloys*, 1940, **11**, 56.
44. A. D. Melaven, *Ind. Eng. Chem., Analyt.*, 1930, **2**, 180.
45. S. E. Wiberley and L. G. Bassett, *Analyt. Chem.*, 1949, **21**, 609.
46. J. J. Lingane, *J. Amer. Chem. Soc.*, 1945, **67**, 1916.
47. R. C. Chirnside, L. A. Dauncey and P. M. C. Proffitt, *Analyst*, 1940, **65**, 446.
48. H. O. Johnson, J. R. Weaver and L. Lykken, *Ind. Eng. Chem., Analyt.*, 1947, **19**, 481.
49. T. D. Parks, H. O. Johnson and L. Lykken, *Analyt. Chem.*, 1948, **20**, 148.
50. J. Maxwell and R. P. Graham, *Chem. Rev.*, 1950, **46**, 471.
51. S. E. Q. Ashley, *Analyt. Chem.*, 1952, **24**, 92.

52. D. D. DeFord, *Analyt. Chem.*, 1956, **28**, 660.
53. J. Rynasiewicz, *Analyt. Chem.*, 1949, **21**, 756.
54. R. B. Hahn, *Analyt. Chem.*, 1953, **25**, 1405.
55. R. J. Hynek and L. J. Wrangell, *Analyt. Chem.*, 1956, **28**, 1520.
56. B. Bagshawe, *J. Iron and Steel Inst.* (*London*), 1954, **176**, 29.
57. E. I. Onstott, *J. Amer. Chem. Soc.*, 1955, **77**, 2129.
58. *Idem, ibid.*, 1956, **78**, 2070.
59. A. Lassieur, *op. cit.*, p. 95.
60. S. Torrance, *Analyst*, 1937, **62**, 719.
61. S. Torrance, *Analyst*, 1938, **63**, 488.
62. A. Lassieur, *op. cit.*, p. 201.
63. J. G. Fife and S. Torrance, *Analyst*, 1937, **62**, 29.
64. E. M. Collin, *Analyst*, 1929, **54**, 654.
65. F. G. Kny-Jones, *Analyst*, 1939, **64**, 172.
66. S. Torrance, *Analyst*, 1939, **64**, 109.
67. H. J. S. Sand, *Analyst*, 1929, **54**, 279.
68. S. Torrance, *Analyst*, 1938, **63**, 489.
69. H. J. S. Sand, *op. cit.* Vol. 2, p. 73.
70. *Idem, Analyst*, 1929, **54**, 279.
71. S. Torrance, *Analyst*, 1939, **64**, 109.
72. G. Norwitz, *Analyt. Chim. Acta*, 1951, **5**, 518.
73. H. W. Miller and R. J. Brouns, *Analyt. Chem.*, 1952, **24**, 536.
74. A. J. Lindsey, *Analyst*, 1948, **73**, 99.
75. A. J. Lindsey and H. J. S. Sand, *Analyst*, 1935, **60**, 739.
76. A. J. Lindsey, *Analyst*, 1938, **63**, 159.
77. A. J. Lindsey and E. A. Tucker, *Analyt. Chim. Acta*, 1954, **11**, 149.
78. *Idem, ibid.*, 1954, **11**, 260.
79. A. J. Lindsey, *Analyst*, 1935, **60**, 744.
80. British Standards Institution, B.S. 1428, Part II, 1954.
81. A. J. Lindsey, *Analyst*, 1948, **73**, 67.
82. B. L. Clarke and H. W. Hermance, *J. Amer. Chem. Soc.*, 1932, **54**, 877.
83. *Idem, Mikrochemie*, 1936, **20**, 126.
84. F. Hernler and R. Pfeningberger, *ibid.*, 1936, Molisch Festschrift, 218.
85. M. Haissinsky, *Bull. Soc. roy. Sci. Liège*, 1951, **20**, 591.
86. *Idem, Experientia*, 1952, **8**, 125.
87. A. D. R. Harrison, A. J. Lindsey and R. Phillips, *Analyt. Chim. Acta*, 1955, **13**, 459.
88. H. J. S. Sand, *Analyst*, 1930, **55**, 309.
89. E. M. Collin, *Analyst*, 1930, **55**, 312.
90. J. G. Fife, *Analyst*, 1936, **61**, 681.
91. P. Ippoliti and E. Scarano, *Ann. Chim.* (*Italy*), 1955, **45**, 492.
92. P. Ippoliti, L. Mercantini and E. Scarano, *Ricerca sci.*, 1955, **25**, 2838.
93. P. Ippoliti and A. Burrati, *Alluminio*, 1956, **25**, 231.
94. J. Guzman and A. Celsi, *Anales Fac. Farm. Bioquim.*, 1938, **9**, 43.
95. J. G. Fife, Ph. D. Thesis, University of London, 1941.
96. E. M. Collin, *Analyst*, 1930, **55**, 312.
97. *Idem, ibid.*, 496.
98. J. G. Fife and S. Torrance, *Analyst*, 1937, **62**, 29.
99. D. L. Carpenter and A. D. Hopkins, *Analyst*, 1952, **77**, 86.
100. E. M. Collin, *Analyst*, 1930, **55**, 680.
101. J. G. Fife, *Analyst*, 1936, **61**, 681.
102. A.S.T.M., Special report No. 98, *Rapid Methods for the Identification of Metals.*
103. H. W. Hermance and H. V. Wadlow, *Physical Methods of Chemical Analysis* W. G. Berl, Ed. (Academic Press, New York, 1951). Vol. II, pp. 156–228.
104. P. R. Monk, *Analyst*, 1953, **78**, 141.
105. H. W. Hermance, *Bell Lab. Rec.*, 1940, **18**, 269.

106. J. A. CALAMARI, *Ind. Eng. Chem., Analyt.*, 1941, **13**, 19.
107. J. A. CALAMARI, R. HUBATH and P. B. ROTH, *Ind. Eng. Chem., Analyt.*, 1942, **14**, 535.
108. G. C. CLARK and E. E. HALE, *Analyst*, 1953, **78**, 145.
109. W. E. SHAW and E. T. MOORE, *Analyt. Chem.*, 1947, **19**, 777.
110. M. KRONSTEIN, M. M. WARD and R. ROPER, *Ind. Eng. Chem.*, 1950, **42**, 1568.
111. BAIRD and TATLOCK (LONDON) LTD., *Instruction Booklet for Electro-spot test Apparatus*. St. Cross Street, Hatton Garden, London, E.C.1.

Chapter III

Potentiometric Titrations

DONALD G. DAVIS

1. Introduction

The measurement of the potential of an appropriate indicator electrode has long been used as a method of detecting the end points of numerous titrations. The potentiometric method is especially versatile, because indicator electrodes suitable for the study of almost every reaction used in titrimetry are now available. The indicator electrodes most often used are the glass electrode and the platinum electrode which, when coupled with appropriate reference electrodes, are used to detect the end points of acid-base and oxidation-reduction titrations respectively. Indicator electrodes made of the metal whose ions form insoluble salts with the anion involved in the titration reaction are generally used to monitor precipitation reactions. A large number of other, sometimes complicated, electrodes have been studied for various specific purposes.

Potentiometric titrations are particularly popular, because they not only establish the equivalence points of many reactions but also provide information about the concentration of one or more of the reactants throughout the titration; possibilities are thus opened up for a reasonably complete understanding of a whole reaction. Such information may not be needed in well studied and much used titrations, but it is very necessary when new reactions or unusual samples are under investigation. Indeed, every new visual indicator that is proposed for general use should be, and usually is, studied potentiometrically in order that its colour characteristics may be exactly established.

Another reason why potentiometric titrations have found so much favour among analytical chemists is that the necessary apparatus is generally inexpensive, reliable, and readily available in most laboratories; in addition, it is generally easy to interpret potentiometric titration curves with a minimum of mathematical effort.

References p. 165

2. Electrode Potentials

A galvanic cell is a system in which the energy produced or consumed by chemical reactions is converted to electrical energy. Many factors influence the value of the e.m.f. of a galvanic cell, including temperature, pressure, and the concentrations of the substances taking part in the electrode reactions.

A typical galvanic cell is formed by the union of two half-cells: for instance, a copper metal electrode immersed in a copper sulphate solution and a zinc metal electrode immersed in a zinc solution each form a half-cell; if these two half-cells are joined in such a way that the zinc sulphate and the copper sulphate solutions are in contact, but prevented from mixing by a porous diaphragm, the galvanic cell so formed may be represented by

$$(-)Zn/ZnSO_4(C_1)//CuSO_4(C_2)/Cu(+). \quad (1)$$

The single oblique strokes in this representation indicate phase boundaries. The double oblique stroke shows a 'liquid junction'. The relative signs of the two electrodes are represented by (+) and (—). The concentrations of the solutions in each half-cell are (C_1) and (C_2).

By convention, the electrode on the left is considered to be the electrode at which an oxidation takes place (the anode) and the right hand electrode is therefore the cathode, where a reduction takes place. The half reactions of galvanic cell (1) will be:

$$\text{at the zinc electrode: } Zn = Zn^{2+} + 2e$$
$$\text{at the copper electrode: } Cu^{2+} + 2e = Cu.$$

The complete cell reaction will be the sum of these half-reactions:

$$Cu^{2+} + Zn = Zn^{2+} + Cu.$$

If the reaction is spontaneous (as it is in this case), electrons will be liberated at the left (anode) and consumed at the right (cathode). Thus, electrons will flow from left to right, which is considered to be positive current, and the e.m.f. of the cell is said to be positive.

The e.m.f. of the cell is the sum of three potentials, one across each of the electrode/solution boundaries and the third across the junction between the two different solutions.

$$E_{cell} = E_{Zn/Zn^{2+}} + E_j + E_{Cu^{2+}/Cu}. \quad (2)$$

The subscripts refer to the direction of the reaction at each electrode, and E_j is the liquid junction potential. The electrode potentials in equation (2) are given by the Nernst equation. For instance, for the copper electrode

$$E_{Cu^{2+}/Cu} = E^0_{Cu^{2+}/Cu} - \frac{RT}{nF} \ln \frac{a_{Cu}}{a_{Cu^{2+}}}, \tag{3}$$

where R is the gas constant (8.316 volt coulombs/degree), T is the absolute temperature, n is the number of electrons taking part in the electrode reaction, and F is the faraday (96,493 coulombs); $E^0_{Cu^{2+}/Cu}$ is the standard potential of the Cu^{2+}/Cu electrode as defined by the cell

$$\text{Pt. } H_2 \text{ (1 atm)}/H^+ (a = 1), Cu^{2+} (a = 1)/Cu. \tag{4}$$

The potential of the left hand electrode (normal hydrogen electrode) is defined as being equal to zero at all temperatures.

The activities of cupric ion and copper metal are designated by $a_{Cu^{2+}}$ and a_{Cu} respectively; the activity of a solid metal is taken as unity. As information about activity coefficients is rarely available, equation (3) is generally written

$$E_{Cu^{2+}/Cu} = E^0_{Cu^{2+}/Cu} + \frac{0.05915}{2} \log_{10} [Cu^{2+}] \tag{5}$$

at 25° after the various numerical values mentioned above have been introduced.

The value of E^0 may be obtained from a table of standard potentials such as those given by Latimer[1] shown in Table III, 3 (p. 83). Care must be taken in the use of various standard potentials because of difficulties with signs. Latimer's original potentials are all written as oxidations, which is contrary to the conventions decided upon at the 1953 meeting of the International Union of Pure and Applied Chemistry. Lingane[2] has discussed sign conventions in a lucid and rational manner and his work should be consulted if difficulties arise.

The potential of the cell represented on p. 66 may be found by combining equation (5) with a similar equation for the zinc electrode and the liquid junction potential (see equation 2). Liquid junction potentials can be assigned in a number of cases[2], but are generally unknown and constitute a possible source of difficulty. Generally, the value of E_j can be made very small by the use of an appropriate salt bridge and, because of this, is generally neglected in potentiometric titration work.

Equation (5) relates potential to concentration for the specific case of the copper electrode. Equations of this general type will be used to interpret potentiometric titration curves and to form the basis for the calculation of end-point potentials.

3. Theoretical Considerations

(a) Precipitation Titrations

A precipitation titration that involves insoluble salts of metals such as mercury, silver, lead and copper may be followed potentiometrically. The indicator electrode may be made of the metal involved in the reaction (a silver electrode for the titration of halides for instance), or may be an electrode whose potential is governed by the concentration of the anion being precipitated, such as the $Pt/Fe(CN)_6^{3-}, Fe(CN)_6^{4-}$ electrode, which could be used to detect the end point when zinc is precipitated with ferrocyanide ion; a small amount of ferricyanide is added to the solution to establish the potential.

The magnitude of the potential change at the end point depends on the solubility of the substance being precipitated as well as on the concentration involved. As an example, the titration of chloride ion with a standard solution of silver nitrate using a silver metal indicator electrode may be considered. The other electrode used to complete the cell is unimportant, provided that it is a true reference electrode, *i.e.* that it maintains a constant potential. In this discussion, it will be assumed that the normal hydrogen electrode (N.H.E.) is used; although this is seldom done in practice, the assumption is convenient because standard potentials may be used directly.

The potential of the silver electrode will be governed by the appropriate Nernst equation:

$$E_{Ag^+/Ag} = E^0_{Ag^+/Ag} + \frac{0.05915}{1} \log_{10} [Ag^+] \tag{6}$$

As soon as enough silver nitrate solution to precipitate some silver chloride has been added, the following equilibrium is established,

$$AgCl \rightleftharpoons Ag^+ + Cl^-, \tag{7}$$

the equilibrium constant of which is

$$K_{AgCl} = [Ag^+] [Cl^-] \simeq 10^{-10}. \tag{8}$$

If $0.1N$ sodium chloride is titrated with $0.1N$ silver nitrate, the silver ion concentration may be considered to be $10^{-9}N$ as soon as the first few drops of silver nitrate have been added. Equation (6) can be used to calculate the indicator electrode potential.

$$E_{Ag^+/Ag} = 0.80 \text{ v} + 0.059 \log_{10} 10^{-9} = 0.27 \text{ v}. \tag{9}$$

Likewise, half-way through the titration when the chloride ion concentration has been reduced to $0.033N$,

$$E_{Ag^+/Ag} = 0.80 \text{ v} + 0.059 \log (3 \times 10^{-9}) = 0.30 \text{ v}.$$

At the equivalence point, $[Ag^+] = [Cl^-] = 10^{-5}N$

$$E_{Ag^+/Ag} = 0.80 \text{ v} + 0.059 \log_{10} 10^{-5} = 0.50 \text{ v}.$$

Thus, by the use of the Nernst equation, the complete titration curve may be constructed as shown in Fig. III, 1.

These rather rough calculations were based on the assumption that only a negligible amount of chloride ion is contributed by the dissolution of silver chloride, even just before the end point; if this assumption is not made, the calculation becomes more involved and a quadratic equation is obtained. (For a discussion involving this more exact treatment, consult Reference 2, p. 119).

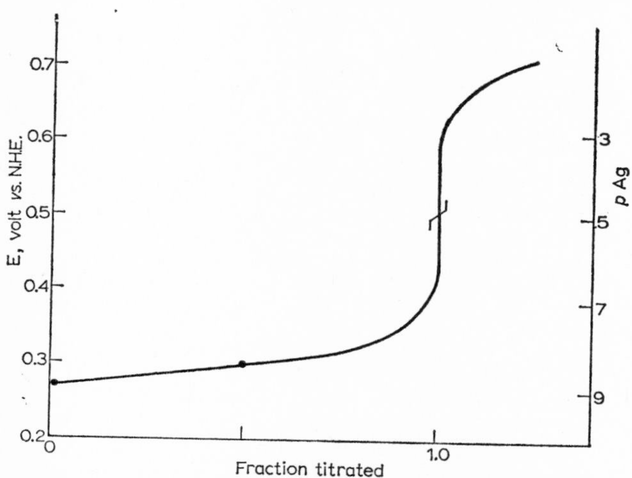

Fig. III, 1. Potentiometric titration of 0.1N chloride with 0.1N silver.

Neglecting any liquid-junction potential, the end-point potential (E_{ep}), may also be directly calculated from the following equation,

$$E_{ep} = E^0{}_{Ag^+/Ag} - \frac{0.05915}{2} pK_{AgX}, \qquad (10)$$

where pK_{AgX} is the negative logarithm of the solubility product of any silver halide, AgX. Knowledge of the value of the end-point potential is necessary for the application of certain specific potentiometric titration methods (see Section 4); for instance, the equivalence point may be located by adding reagent until the calculated equivalence-point potential is exactly obtained.

References p. 165

It is usually necessary to control experimental conditions carefully to be able to obtain the calculated value of the equivalence-point potential. Generally, the equivalence point is found by differentiation: if the change in potential is expressed as a function of the volume of reagent added, it can be seen (Table III, 1) that the maximum value of dE/dV is obtained at the exact equivalence point. (See also Section 4c, p. 102.)

As it is simple to construct a table such as this directly from the titration figures, provided that the titrant is added in small equal increments, the equivalence point is usually found in this way.

TABLE III, 1

TITRATION OF 100 ml OF 0.1*N* CHLORIDE ION WITH 0.1*N* SILVER ION

0.1*N* Ag^+,	p*Ag*	*E vs. N.H.E.*	dE/dV
99.7 ml	6.18	+0.435 mv	
			11
99.8	6.00	+0.446	
			18
99.9	5.70	+0.464	
			41
100.0	5.00	+0.505	
			41
100.1	4.30	+0.546	
			18
100.2	4.00	+0.564	
			11
100.3	3.82	+0.575	

In the titration of any substance that forms a precipitate containing an unequal number of anions and cations, the titration curve is not symmetrical and dE/dV appears when there is a small excess of either anions or cations. Kolthoff and Furman[3] give equations that can be used to calculate the titration error, if the maximum value of dE/dV is assumed to occur at the equivalence point of unsymmetrical titrations. As the error depends on the concentration of the solution titrated as well as on the magnitude of the solubility product, more exact results can probably be obtained by titrating to a previously calculated or experimentally determined equivalence-point potential. In practical work, the maximum value of dE/dV is usually taken as the end point in any case, and experimental conditions are arranged so that only tolerable errors will be incurred. For instance, concentrated solutions may be used, or some solvent such as acetone may be added to reduce the solubility of the precipitate.

The end-point potential of a titration of an unsymmetrical nature may be calculated as illustrated by the following example: for silver oxalate,

$$K_{Ag_2Ox} = [Ag^+]^2\,[Ox^{2-}]; \qquad (11)$$

at the equivalence point,

$$\tfrac{1}{2}[Ag^+] = [Ox^{2-}] \qquad (12)$$

$$\frac{[Ag^+]^3}{2} = K_{Ag_2Ox} \qquad (13)$$

$$[Ag^+] = \sqrt[3]{2K_{Ag_2Ox}} \qquad (14)$$

$$E_{ep} = E^0{}_{Ag^+/Ag} + 0.05915 \log_{10} (2K_{Ag_2Ox})^{1/3} \qquad (15)$$

Table III, 2 gives the equivalence-point potentials and solubility products of some commonly precipitated silver salts; the data are taken from Latimer's "Oxidation Potentials" (see Reference 1).

TABLE III, 2

SOLUBILITY PRODUCTS AND EQUIVALENCE-POINT POTENTIALS OF SOME INSOLUBLE SILVER SALTS

Salt	*K*	E_{ep}
AgI	8.5×10^{-17}	+0.325
AgCN	7.0×10^{-15}	+0.38
AgBr	3.3×10^{-13}	+0.437
AgCNS	1.0×10^{-12}	+0.45
AgCl	1.72×10^{-10}	+0.512
Ag_2CrO_4	1.1×10^{-12}	+0.570
Ag_2O	2.0×10^{-8}	+0.573
$AgIO_4$	3.04×10^{-8}	+0.578
$Ag_2C_2O_4$	1.1×10^{-11}	+0.59
$AgBrO_3$	5.20×10^{-5}	+0.674

One of the great advantages of potentiometric titrations is that several components may be titrated in the same solution without the possibility of indicators interfering with each other; for instance, iodide and bromide may be titrated together, giving a titration curve such as is shown in Fig. III, 2. If 50 ml of a solution 0.1*N* in iodide and 0.1*N* in bromide is titrated with 0.1*N* silver nitrate, the first precipitate to form will be silver iodide. Silver iodide will continue to be precipitated alone until the silver ion concentration reaches a value of

$$\frac{K_{AgBr}}{[Br^-]} \quad \text{or } 3.3 \times 10^{-12},$$

at which silver bromide will just begin to form. At this point (A in Fig. III, 2),

$$[I^-] = \frac{K_{AgI}}{[Ag^+]} = \frac{8.5 \times 10^{-17}}{3.3 \times 10^{-12}} = 2.6 \times 10^{-5}N.$$

In other words, if the end point of the iodide titration is taken at point A, only 0.026% of the original iodide ion will remain unprecipitated. If 0.001N iodide were titrated in the presence of 0.1N bromide, the error would be increased to −2.6%.

Mixtures of bromide and chloride can also be titrated together, and the error in the bromide determination for 0.1N solutions will be about 0.2%.

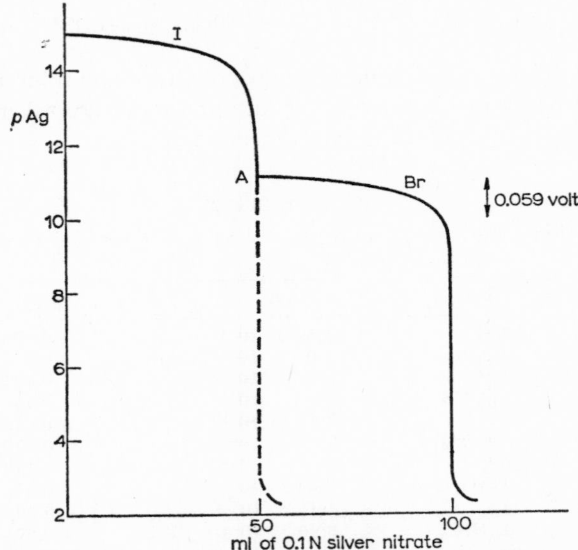

Fig. III, 2. Potentiometric titration of 50 ml of solution, 0.1N in both I^- and Br^-, with silver nitrate.

In actual practice, mixtures of halides cannot be titrated as accurately as these calculations appear to indicate. Silver halides form solid solutions rather easily, so that there is a fairly large amount of co-precipitation of the more soluble silver halide with the less soluble. Errors of about 2% or more are usually obtained in the iodide and bromide end points with mixtures of all three halides; the sum of all three can, however, be determined very accurately. (For further information see Section 5*b*.)

(b) Acid-Base Titrations

A number of different indicator electrodes may be employed in order to follow the progress of acid-base reactions, the most commonly used being the glass electrode, but, regardless of which electrode is chosen, the potential of the electrode is given by an equation of the form

$$E = E^1 + 0.0001984\ T\text{pH},$$

where E^1 is a constant potential depending on the experimental arrangement, the liquid junction potential(s) and the reference electrode, and T is the absolute temperature. The titration curve obtained by plotting E against volume of titrant added is of the same form as shown in Fig. III, 1; the equivalence point is generally indicated by a maximum value of dE/dV.

(*i*) *Strong Acids and Bases*

For the titration of a strong acid with a strong base, the titration reaction is

$$H^+ + OH^- = H_2O, \tag{16}$$

$$K = \frac{1}{K_w} = 10^{+14}.$$

At the equivalence point for a temperature of 25°,

$$[H^+] = [OH^-] = 10^{-7};\ \text{pH} = 7. \tag{17}$$

As strong acids are essentially completely dissociated, the hydrogen ion concentration will be equivalent to the concentration of acid (plus a small contribution from the dissociation of water, which is generally neglected except in the most exact work). The pH before the end point may be calculated from the amount of strong acid taken, the total volume, and the amount of base added.

Beyond the equivalence point, the pH is conveniently found from a knowledge of the amount of excess of hydroxide ion added.

The pH at the equivalence point of the titration of a strong acid with a strong base is always theoretically 7.00 at 25°, regardless of the concentration of the acid. This is not true for weak acids, in which case the pH of the equivalence point decreases as the concentration decreases.

In practical work, the maximum value of dE/dV is frequently found below pH 7, because of the presence of impurities such as dissolved carbon dioxide. If the carbon dioxide concentration is rather large, two inflections can be noted in the titration curve. For exact work, this difficulty may be avoided by working with carbonate-free materials.

(*ii*) *Weak Acids and Bases*

The titration reaction of the neutralisation of a weak acid, HA is

References p. 165

$$HA + OH^- = A^- + H_2O \tag{18}$$

$$K_{eq} = \frac{[A^-]}{[HA]\,[OH^-]} = \frac{K_a}{K_w}, \tag{19}$$

where

$$K_a = \frac{[H^+]\,[A^-]}{[HA]} \tag{19a}$$

At the equivalence point, reaction (18) has not gone to completion, indicating that the pH will be somewhat greater than 7.

At the start of the titration, the pH is given by

$$pH = \tfrac{1}{2}pK_a - \tfrac{1}{2}\log[HA] \tag{20}$$

(For a more exact treatment see Vol. I B, Chapter VII).

Equation (19) or (19a) may be used to calculate the pH between the start of the titration and the equivalence point. At the equivalence point, $[OH^-] = [HA]$ by definition, so that equation (19) may be rearranged to:

$$[OH^-] = \sqrt{\frac{K_w}{K_a}[A^-]}, \tag{21}$$

or

$$pOH = \tfrac{1}{2}(pA + pK_w - pK_a)$$
$$pH = pK_w - pOH = \tfrac{1}{2}(pK_w - pA + pK_a) \tag{22}$$

Beyond the equivalence point, the excess of hydroxide ion will drive reaction (18) essentially to completion, so that the pH may be calculated on the assumption that excess of base is being added to pure water.

The titration curve for a weak base with a strong acid may be calculated from formulae analogous to those given above. The maximum value of dE/dV, (dpH/dV), is generally taken as the end point, even though the titration curve of a weak acid with a strong base is unsymmetrical. This practice results in measurable errors only when the concentration of the acid or its K_a value is very small. Practically, the limit is reached[3] when K_a becomes as small as 10^{-8} at a concentration level of $0.1N$. For each shift in the concentration by a power of 10, the K_a value must be greater by a power of 10 for good results to be obtained. It has been calculated[4] that an error of only 0.3% will be found between the maximum value of dE/dV and the equivalence point, if the product of the concentration and K_a is equal to 10^{-11}; this allows no experimental error, which, for a very weak acid, could markedly effect the exact position of the maximum value of dE/dV, as the maximum would not be sharp.

Figure III, 3 shows how the variation of K_a effects the form of the titration curve.

A weak acid may also be titrated with a weak base, but this procedure does not result in a titration curve with a marked inflection, owing to the fact that hydrolysis is important on both sides of the equivalence point.

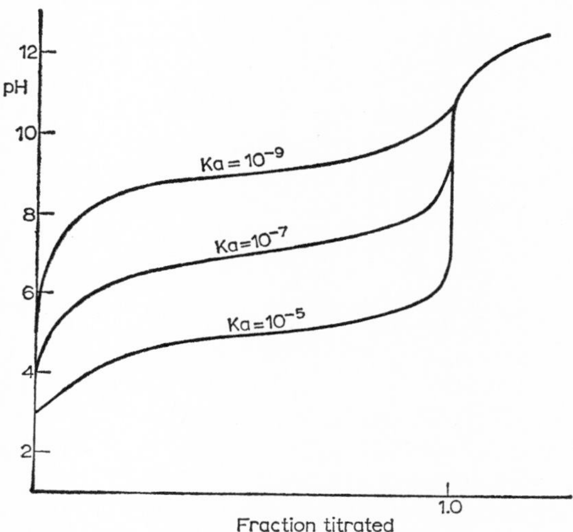

Fig. III, 3. Titration curves of various weak acids with 0.1*N* sodium hydroxide.

iii) Polyprotic Acids and Mixtures of Weak Acids

The neutralisation curve for a polyprotic acid or a mixture of weak acids will show two inflection points, provided that the acids differ sufficiently in strength: the first occurs after the neutralisation of the stronger acid and the second after the neutralisation of the weaker.

In general, the titration curve may be calculated on the assumption that only one acid need be considered at a time. Thus, during the first part of the titration curve almost up to the first inflection, only the stronger acid need be considered, and the pH may be calculated as indicated in Section 2. Likewise, after the first inflection only the ionisation *etc.* of the weak acid need be considered; this includes the calculation of the pH at, and beyond, the second inflection.

The pH at the first equivalence point may be calculated as follows, provided that the value of K_1 is much smaller than the concentration of the predominant acid species (HA^-):

$$\mathrm{pH} = \tfrac{1}{2}(\mathrm{p}K_1 + \mathrm{p}K_2). \qquad (23)$$

References p. 165

For equation (23) to be used, the acid in question must be polyprotic or, with a mixture of two weak acids, their concentrations must be equal. If their concentrations are not equal, as would be the usual case, equation (23) must be modified to:

$$\mathrm{pH} = \tfrac{1}{2}(\mathrm{p}K_1 + \mathrm{p}K_2) - \tfrac{1}{2}\log C_1/C_2, \tag{24}$$

where C_1 and C_2 are the concentrations of the stronger and weaker acids respectively.

As might be expected, there are certain limitations to the use of these formulae and to the ability of the potentiometric method to show the inflection points involved. No inflection at all occurs with two acids of the same concentration when the stronger acid is no more than 100 times stronger than the weaker. It is generally considered[3] that, if an accuracy of 0.5% is desired, the difference in the dissociation constants should be 10^5 at the same approximate concentration.

(c) Complexometric Titrations

Complexometric titrations can be followed with an electrode of the metal whose ion is involved in complex formation. For instance, a silver electrode may be used to follow the titration of cyanide ion with a standard silver solution. The potential of the silver electrode may be expressed by equation (6) (p. 68). The silver ion concentration will be governed by the equilibrium constant for the complex formation reaction

$$Ag^+ + 2CN^- = Ag(CN)_2^-, \tag{25}$$

$$K_c = \frac{[Ag^+]\,[CN^-]^2}{[Ag(CN)_2^-]}. \tag{26}$$

The titration curve may be exactly calculated from equations (6) and (26) together with the known concentrations of cyanide and silver. At the equivalence point,

$$2\,[Ag^+] = [CN^-], \tag{27}$$

so that equation (26) becomes

$$[Ag^+] = \sqrt[3]{\frac{K_c}{4}\,[Ag(CN)_2^-]}. \tag{28}$$

In this case, solid silver cyanide begins to be precipitated soon after the equivalence point. A small further addition of silver does not change the concentration of the complex nor of the silver ion to any extent, so that the titration curve has an almost horizontal portion shortly after the equivalence point (Fig. III, 4).

In many complexometric reactions, the situation cannot be handled so easily because more than one complex is formed, as occurs in the reaction of mercuric ion and cyanide:

$$Hg^{2+} + 3CN^- = Hg(CN)_3^-$$

$$Hg^{2+} + 4CN^- = Hg(CN)_4^{2-}$$

The dissociation constants of both complex ions must be considered for the calculation of the titration curve of such a reaction. Often, when more than

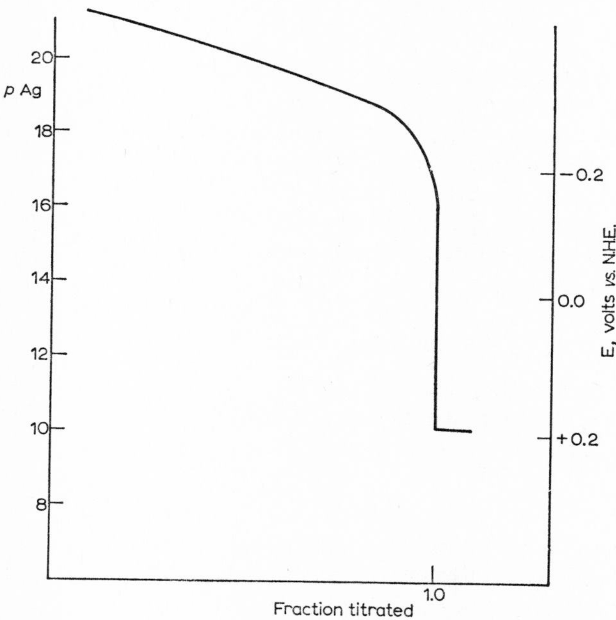

Fig. III, 4. Titration of 0.1*N* cyanide with 0.1*N* silver solution.

one complex species is formed, the d*E*/d*V* maximum is not pronounced enough to allow accurate titrations to be performed. It happens, however, that the mercuric cyanide complexes are so strong that good results may be obtained, despite the fact that two different complex ions may be produced[5].

Since 1945, complexometric titrations have become widely used because of the discovery of the analytical importance of metal chelating agents such as ethylenediaminetetra-acetic acid (EDTA), which is frequently denoted

References p. 165

by the symbol Y. Potentiometric end-point detection has been successfully applied to EDTA titrations, particularly by Reilley[6]. He was able to follow many EDTA titrations by relying on the interaction between a metal ion, M^{2+}, and a small amount of mercury EDTA complex, HgY^{2-}

$$M^{2+} + HgY^{2-} = MY^{2-} + Hg^{2+}. \quad (29)$$

As the potential of a mercury electrode depends on the concentration of Hg^{2+},

$$E = E^0_{Hg/Hg^{2+}} + \frac{0.05915}{2} \log_{10} [Hg^{2+}], \quad (30)$$

the extent to which reaction (29) proceeds could be followed potentiometrically.

Complexometric titrations can often be monitored by measuring the effect that the complexing agent has on some oxidation-reduction couple such as Fe^{3+}/Fe^{2+}: for instance, the standard potential of the FeY^{-}/Fe^{2+} couple is greatly different than that of the Fe^{3+}/Fe^{2+} couple, and a substantial inflection occurs in the titration curve of ferric iron with EDTA when a platinum electrode is used[7]. The theory of such titrations is similar to that for standard oxidation-reduction titrations and section *d* should be consulted for further information.

(d) Oxidation-Reduction Titrations

Oxidation-reduction titrations, unlike the three cases discussed above (sections *a-c*), involve the transfer of electrons from the substance being oxidised to the substance being reduced: for instance,

$$Ce^{4+} + Fe^{2+} = Fe^{3+} + Ce^{3+}. \quad (31)$$

It is generally considered that such a reaction consists essentially of two half-reactions, whose standard potentials may be used to calculate the standard potential of the net reaction, (31).

$$Fe^{2+} = Fe^{3+} + e, \quad E^0 = -0.76 \text{ v}; \quad (32)$$

$$Ce^{4+} + e = Ce^{3+}, \quad E^0 = +1.61 \text{ v}; \quad (33)$$

$$Ce^{4+} + Fe^{2+} = Fe^{3+} + Ce^{3+}, \quad E^0 = +0.85 \text{ v}. \quad (34)$$

The equilibrium constant of any reaction may be calculated from the following formula:

$$E^0 = \frac{0.05915}{n} \log_{10} K,$$

where n is best defined as the number of equivalents of electricity associated with one molar unit of reaction. For the reduction of ceric by ferrous ions, (see above) $K = 2.5 \times 10^{14}$.

If an acidic ferrous solution is titrated with a standard ceric solution at 25°, the potential (relative to the N.H.E.) of a platinum electrode in contact with the solution will be given by either of the following equations:

$$E = E^0{}_{Ce^{4+}/Ce^{3+}} - \frac{0.0591}{1} \log_{10} \frac{[Ce^{3+}]}{[Ce^{4+}]}, \tag{35}$$

$$E = E^0{}_{Fe^{3+}/Fe^{2+}} - \frac{0.0591}{1} \log_{10} \frac{[Fe^{2+}]}{[Fe^{3+}]}. \tag{36}$$

It would be more convenient to use equation (36) before the equivalence point, as the right-hand term of this equation could be easily found from the known extent of the titration. Equation (35) could also be used, but the $[Ce^{3+}]:[Ce^{4+}]$ ratio would have to be calculated by means of the equilibrium constant. Beyond the equivalence point, calculations are most easily made by means of equation (35).

It should be mentioned that equations (35) and (36) may only be applied when equilibrium has been attained; this usually occurs rapidly, however, and causes no trouble in most practical cases.

From equation (36), it is evident that the potential at the start of the titration should be $-\infty$, because Fe^{2+} is the only ion present and there are no Fe^{3+} ions. In practice, however, a pure ferrous solution can never be prepared, because ferrous ions can be oxidised by hydrogen ions, if only to a slight extent.

At the mid-point of the titration, where $[Fe^{2+}] = [Fe^{3+}]$, equation (36) becomes:

$$E = E^0{}_{Fe^{3+}/Fe^{2+}}. \tag{37}$$

At the equivalence point, the concentration of unchanged ferrous ion will be equal to the concentration of unchanged ceric ion. Likewise, the concentration of cerous ion will be equal to the concentration of ferric ion.

Therefore it is true that

$$\frac{[Fe^{2+}]}{[Fe^{3+}]} = \frac{[Ce^{4+}]}{[Ce^{3+}]}, \tag{38}$$

and, as

$$K = \frac{[Fe^{3+}]\,[Ce^{3+}]}{[Fe^{2+}]\,[Ce^{4+}]}, \tag{39}$$

References p. 165

it is evident that, at the equivalence point,

$$\frac{[Fe^{3+}]}{[Fe^{2+}]} = \frac{[Ce^{3+}]}{[Ce^{4+}]} = \sqrt{K}. \tag{40}$$

Combining equations (35) and (36) with (40) it is found that,

$$E_{ep} = E^0_{Fe^{3+}/Fe^{2+}} + \frac{0.0591}{2} \log_{10} K; \tag{41}$$

$$E_{ep} = E^0_{Ce^{4+}/Ce^{3+}} - \frac{0.0591}{2} \log_{10} K. \tag{42}$$

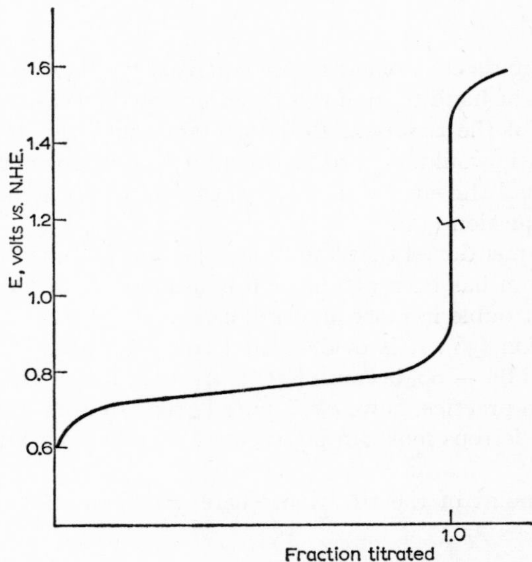

Fig. III, 5. Titration of 0.1*M* ferrous solution with 0.1*M* ceric solution.

By adding equation (41) to equation (42) the following expression is obtained for the potential at the equivalence point:

$$E_{ep} = \frac{E^0_{Fe^{3+}/Fe^{2+}} + E^0_{Ce^{4+}/Ce^{3+}}}{2}. \tag{43}$$

It should be noticed that the equivalence-point potential is completely independent of the concentrations of the reactants and the products.

Fig. III, 5 shows a theoretical titration curve calculated according to the procedure given above. Equation (43) is solved as follows:

$$E_{ep} = \frac{0.76 + 1.61}{2} = 1.19 \text{ V}$$

The titration curve is symmetrical about the equivalence point, so that the maximum value of dE/dV is found exactly at the equivalence point; this is true of all reactions in which the oxidant and the reductant react in a 1:1 ratio.

In a more complicated case, such as

$$a\text{Ox}_1 + b\text{Red}_2 = a\text{Red}_1 + b\text{Ox}_2, \quad (44)$$

the coefficients and the amount of electricity involved per molar unit must be considered in the calculation of the titration curve. Nernst equations for the appropriate half-reaction should still be used, but the potential at the equivalence point must be calculated by

$$E_{ep} = \frac{bE_1 + aE_2}{a + b}; \quad (45)$$

for instance, if ferrous ion is oxidised by permanganate ion, equation (45) becomes

$$E_{ep} = \frac{5 \times 1.52 + 0.76}{6} = 1.39 \text{ V},$$

if $[H^+] = 1$.

Here, the equivalence-point potential is no longer the mean of the two values of E^0, but lies much closer to the permanganate-manganous standard potential. Because of this, the curve is not symmetrical about the equivalence point, and the maximum value of dE/dV is actually slightly in advance of the equivalence point. The rate of change of the potential at the equivalence point is, however, so great that no significant error is introduced by using the maximum of dE/dV as the end point, provided that the concentrations of the solutions involved are not extremely small. The theory of oxidation-reduction systems has been extensively discussed by Nightingale[8].

It should be remembered that the equivalence-point potential of an oxidation-reduction reaction involving hydrogen ions is dependent on the hydrogen ion concentration. Advantage can sometimes be taken of this fact to increase the value of dE/dV at the equivalence point, as the effect of the hydrogen ion is usually greater for one half-reaction than for the other.

When the lower oxidation states of some elements, such as vanadium, are titrated with a strong oxidant, it is often possible to obtain a potentiometric titration curve with several breaks corresponding to stepwise oxida-

References p. 165

tions to successively higher oxidation states; with vanadium, three inflections can be obtained corresponding to

$$V^{2+} = V^{3+} + e, \quad E^0 = +0.25 \text{ v}, \tag{46}$$

$$V^{3+} + H_2O = VO^{2+} + 2H^+ + e, \quad E^0 = -0.314 \text{ v}, \tag{47}$$

$$VO^{2+} + H_2O = VO_2^+ + 2H^+ + e, \quad E^0 = -1.00 \text{ v}, \tag{48}$$

when V^{2+} is titrated with permanganate in sulphuric acid. In order to obtain a distinct break in a stepwise reaction of this type, the potentials of the half-reactions must differ by 0.2 v or more.

Such a titration curve may be calculated in a similar way to a regular oxidation-reduction curve. In this case, however, only the potentials of those couples that are of importance on either side of a specific inflection are used to calculate the equivalence-point potential of that inflection. Thus, the first equivalence-point potential would be

$$E_{ep1} = \frac{E_{V^{3+}/V^{2+}} + E_{VO^{2+}/V^{3+}}}{2},$$

the second

$$E_{ep2} = \frac{E_{VO^{2+}/V^{3+}} + E_{VO_2^+/VO^{2+}}}{2},$$

and the third

$$E_{ep3} = \frac{E_{VO_2^+/VO^{2+}} + 5E_{MnO_4^-/Mn^{2+}},}{6}\,; [H^+] = 1.$$

It is possible also to titrate two or more elements in the same solution, provided that the potentials of the various couples involved differ by several tenths of a volt. For instance, V^{3+} and U^{4+} can be titrated with a strong oxidant such as permanganate, but two inflections are observed rather than three. This is because the oxidation of V^{3+} to V^{4+} takes place at a potential very close to that of the oxidation of U^{4+} to U^{6+}, so that the first observed break corresponds to the titration of both vanadium and uranium. The amount of vanadium can be found, because the oxidation of V^{4+} to V^{5+} yields a second break. Indirectly, the amount of uranium may then be calculated, provided that all vanadium in the solution titrated was originally in the 3+ oxidation state.

Table III, 3 lists the standard potentials of a number of half-reactions that are commonly used in potentiometric titrimetry; these are useful for

TABLE III, 3

STANDARD AND FORMAL REDUCTION POTENTIALS (AT 25°)*

Reaction	E^0 *vs. N.H.E.*
$Zn^{2+} + 2e = Zn$	−0.758 v
$Fe(OH)_3 + e = Fe(OH)_2 + OH^-$	−0.65
$U^{4+} + e = U^{3+}$	−0.61
$Cr^{2+} + 2e = Cr$	−0.60
$S + 2e = S^{2-}$	−0.51
$Fe^{2+} + 2e = Fe$	−0.44
$2H^+ + 2e = H_2$ (pH = 7.0)	−0.414
$Cr^{3+} + e = Cr^{2+}$	−0.41
$Ti^{3+} + e = Ti^{2+}$	−0.37
$Co(CN)_6^{3-} + e = Co(CN)_6^{4-}$	−0.3
$V^{3+} + e = V^{2+}$	−0.255
$SnCl_4^{2-} + 2e = Sn + 4Cl^-$ (1*M* HCl)	−0.19
$AgI + e = Ag + I^-$	−0.151
$Sn^{2+} + 2e = Sn$	−0.136
$Pb^{2+} + 2e = Pb$	−0.126
(1*M* H_2SO_4)	−0.29
(2*M* NaAc)	−0.32
$2H^+ + 2e = H_2$	±0.000
$AgBr + e = Ag + Br^-$	+0.073
$AgSCN + e = Ag + SCN^-$	+0.095
$Co(NH_3)_6^{3+} + e = Co(NH_3)_6^{2+}$	+0.1
$TiO^{2+} + 2H^+ + 2e = Ti^{3+} + H_2O$	+0.1
$TiOCl^+ + 2H^+ + 3Cl^- + e = TiCl_4^- + H_2O$ (1*M* HCl)	−0.09
(6*M* HCl)	+0.24
$Ti^{4+} + e = Ti^{3+}$ (5*M* H_3PO_4)	−0.15
$S + 2H^+ + 2e = H_2S$	+0.141
$Cu^{2+} + e = Cu^+$	+0.153
$Sn^{4+} + 2e = Sn^{2+}$	+0.154
(1*M* HCl)	+0.14
$SO_4^{2-} + 4H^+ + 2e = H_2SO_3 + H_2O$	+0.17
$Co(OH)_3 + e = Co(OH)_2 + OH^-$	+0.17
$AgCl + e = Ag + Cl^-$	+0.222
Saturated calomel electrode	+0.2415
$Hg_2Cl_2 + 2e = 2Hg + 2Cl^-$	+0.2676
$Cu^{2+} + 2e = Cu$	+0.345
$Fe(CN)_6^{3-} + e = Fe(CN)_6^{4-}$	+0.356
(1*M* HCl or $HClO_4$)	+0.71
$UO_2^{2+} + 4H^+ + 2e = U^{4+} + H_2O$	+0.334
$VO^{2+} + 2H^+ + e = V^{3+} + H_2O$	+0.361
$Cu^{2+} + 2Cl^- + e = CuCl_2^-$	+0.46
$I_{2(s)} + 2e = 2I^-$	+0.535

* These potentials are mainly from W. M. Latimer, *Oxidation Potentials* (Prentice Hall, Inc., New York, 2nd Ed., 1952); E. H. Swift, *A System of Chemical Analysis* (W. H. Freeman and Co., San Francisco, 1939); and J. J. Lingane, *Electroanalytical Chemistry* (Interscience, New York, 1958).

TABLE III, 3 (*continued*)

Reaction	E^0 *vs. N.H.E.*
$I_3^- + 2e = 3I^-$	+0.536 v
$H_3AsO_4 + 2H^+ + 2e = HAsO_2 + 2H_2O$	+0.559
$MnO_4^- + e = MnO_4^{2-}$	+0.564
$Hg_2SO_4 + 2e = 2Hg + SO_4^{2-}$	+0.615
$O_2 + 2H^+ + 2e = H_2O_2$	+0.682
$Fe^{3+} + e = Fe^{2+}$	+0.771
($1M$ H_2SO_4)	+0.68
($1M$ H_2SO_4 − $0.5M$ H_3PO_4)	+0.61
$Hg_2^{2+} + 2e = 2Hg$	+0.789
$Ag^+ + e = Ag$	+0.7995
$Cu^{2+} + I^- + e = CuI$	+0.86
$ClO^- + H_2O + 2e = Cl^- + 2OH^-$	+0.89
$2Hg^{2+} + 2e = Hg_2^{2+}$	+0.920
$NO_3^- + 4H^+ + 3e = NO + 2H_2O$	+0.96
$VO_2^+ + 2H^+ + e = VO^{2+} + H_2O$	+1.000
$AuCl_4^- + 3e = Au + 4Cl^-$	+1.00
$2ICl_2^- + 2e = I_2 + 4Cl^-$	+1.06
Br_2(liq.) + 2e = $2Br^-$	+1.065
Br_2(aq) + 2e = $2Br^-$	+1.087
$ClO_4^- + 2H^+ + 2e = ClO_3^- + H_2O$	+1.19
$IO_3^- + 6H^+ + 5e = \frac{1}{2}I_2 + 3H_2O$	+1.20
$O_2 + 4H^+ + 4e = 2H_2O$	+1.229
$MnO_2 + 4H^+ + 2e = Mn^{2+} + 2H_2O$	+1.23
$Tl^{3+} + 2e = Tl^+$	+1.25
($1M$ HCl)	+0.77
$Cr_2O_7^{2-} + 14H^+ + 6e = 2Cr^{3+} + 7H_2O$	+1.33
$Cl_2 + 2e = 2Cl^-$	+1.3595
$PbO_2 + 4H^+ + 2e = Pb^{2+} + 2H_2O$	+1.455
$Au^{3+} + 3e = Au$	+1.50
$MnO_4^- + 8H^+ + 5e = Mn^{2+} + 4H_2O$	+1.51
$Bi_2O_4 + 4H^+ + 2e = 2BiO^+ + 2H_2O$	+1.59
$MnO_4^- + 4H^+ + 3e = MnO_2 + 2H_2O$	+1.695
$Ce^{4+} + e = Ce^{3+}$ ($1M$ $HClO_4$)	+1.70
($1M$ HNO_3)	+1.61
($1M$ H_2SO_4)	+1.44
$H_2O_2 + 2H^+ + 2e = 2H_2O$	+1.77
$Co^{3+} + e = Co^{2+}$	+1.842
$Ag^{2+} + e = Ag^+$ ($4M$ HNO_3)	+1.927
($4M$ $HClO_4$)	+1.970
$S_2O_8^{2-} + 2e = 2SO_4^{2-}$	+2.01
$O_3 + 2H^+ + 2e = O_2 + H_2O$	+2.07

predicting whether distinct inflections are to be expected in any particular titration, as well as in the calculation of equivalence-point potentials. (A more complete list can be found in either reference 1 or 2.)

In actual practice, the use of standard potentials is complicated by the fact that concentrated solutions containing more than one electrolyte are almost always used. Because of this, the activities of the species involved are frequently quite different from their concentrations, and this sometimes makes predictions unreliable. In addition, the actual species involved should be known with some certainty, because the formation of complexes with one or more of the substances involved can cause a substantial variation in the potential of the half-reaction. The formation of complexes can be treated theoretically, provided that the equilibrium constants of the complex-forming reactions are known. Thus, when possible, a separate potential is reported for the equilibrium between such complex ions as ferrocyanide and ferricyanide.

Swift[9] has advocated the use of formal potentials to avoid difficulties due to activities and, in some cases, to complex formation. A formal potential is the experimentally observed potential in a solution containing equal formal concentrations (formula weights per litre of solution) of the oxidised and reduced species and other specified substances at specific concentrations. A number of formal potentials are given in Table III, 3, because they are often useful for the prediction of the results of practical titrations.

Too much faith should not be put in predictions based solely on the potentials of half-reactions, because it has been found in a number of cases that the rates of reaction are such that equilibrium is attained only slowly. If the reaction rate is too slow, a satisfactory titration cannot be carried out, even though the formal potentials are several tenths of a volt apart. This difficulty can sometimes be obviated by heating the solution, but this is not always practicable.

4. Techniques and Apparatus

(a) Requirements for Successful Potentiometric Titrations

Chemical reactions must fulfil certain specific requirements, if they are to be used in any kind of titration. The reaction must proceed in one way only; that is, any side reactions must be of negligible importance. There are reactions that yield several different sets of products, and these can only be used if conditions that yield only a definite ratio of the various sets of products can be established. In certain cases, such as the oxidation of sugars in basic solution with cupric ions, empirical tables can be constructed to take into account the extent of various side reactions. In addition, the reaction should be stoichiometric and must proceed almost completely in one direction. Finally, the rate at which the reaction reaches a state of equilibrium must be relatively rapid, or an impracticable amount of time would be involved in performing a titration.

References p. 165

If these general conditions are fulfilled, a successful potentiometric titration may be carried out, provided that some "indicator" electrode can be found. The potential of the indicator electrode must depend on the concentration of one of the species involved in such a way that a measurable potential change is found at or very near the equivalence point.

Glass, silver, and platinum indicator electrodes are the most generally used for acid-base, precipitation, and oxidation-reduction reactions respectively; these electrodes cannot, however, be used as indicator electrodes in certain reactions: for instance, the silver electrode is only useful for precipitation titrations involving anions that form fairly insoluble silver salts. If lead solutions are to be titrated or used as reagents, a lead electrode would seem to be the logical choice as an indicator electrode, but it does not give very reproducible results, especially at low concentrations. This is fairly typical of non-noble metal electrodes, and of metals such as zinc or cadmium, which also yield poor results. Such electrodes are considered to behave irreversibly because of oxidation, passivation, or the evolution of hydrogen.

In order to make titrations with lead ions possible, an electrode whose potential depends on the anion involved is often used. If a ferrocyanide solution is to be determined with lead, the end point will be indicated by a platinum electrode, provided that a small amount of ferricyanide is added to the solution. In this case, the potential of the platinum electrode depends on the ratio of the concentration of ferrocyanide to that of ferricyanide, because the ferro/ferricyanide couple is a reversible oxidation-reduction system.

When sulphate is determined with lead or barium, neither of the abovementioned types of electrode can be used, because neither lead nor barium is a metal noble enough to be used as a reversible electrode. In addition, sulphate ion does not participate in any reversible oxidation-reduction couples, so that the potential of a platinum electrode cannot be made to depend on the sulphate ion concentration. The precipitation of lead with sulphate can be followed potentiometrically by using a platinum electrode whose potential is established by the ferrocyanide/ferricyanide couple[10]; in fact, this electrode may be considered as a ferricyanide/lead ferrocyanide electrode. Likewise, an iodine/silver iodide electrode can be used for silver ions.

It is evident then that at least one reversible couple participating in, or directly affected by, the titration reaction must be present, if a successful potentiometric titration is to be carried out.

With many oxidation-reduction reactions, difficulties are sometim[illegible] perienced because of irreversibility. For example, the thiosulphate/te[illegible] thionate couple does not establish a reversible potential at a platinum electrode. The reason for this is not well understood, but it is believed that reactions such as the oxidation of thiosulphate take place by way of several intermediate steps. Even so, many irreversible reactions may be used in potentiometric titrations, provided that some reversible couple is present to establish the indicator electrode potential. Thiosulphate may be titrated with iodine, or *vice versa,* and a large potential change is found near the equivalence point. In this case, the indicator electrode potential is established mainly by the iodine/iodide couple. It is not possible, however, to plot the titration curve by the use of the appropriate Nernst equations, and whether such a titration is possible must therefore be determined experimentally.

Because oxidation-reduction reactions take place in steps, especially when the electron change is large, the Nernst equation does not always apply in the expected way: for instance, in the reduction of dichromate, the potential has been found experimentally to be independent of the chromic ion concentration, despite the fact that this concentration appears in the Nernst equation.

Since the earliest measurements of redox (oxidation-reduction) potentials, it has been generally assumed that platinum and other noble metal electrodes are essentially inert, despite the fact that many phenomena that seemed contrary to this assumption were noticed. It has now been established that platinum and gold electrodes become coated with an oxide film when treated with strong oxidants; Kolthoff and Tanaka[11] have conclusively demonstrated the existence of such films by means of current-potential curves. Ross and Shain[12] recorded the slowly drifting potential of a platinum electrode immersed in a number of oxidants and reductants at various concentrations; they found that the time required to establish a steady potential in any specific solution depended on the previous treatment of the electrode (oxidised or reduced), as well as on the nature and concentration of the redox species in solution. Lee, Adams, and Bricker[13] have found the existence of potential regions where the potentials of various noble metal electrodes change only slowly with time.

Anson and Lingane[14] have found chemical evidence for the existence of platinum oxide films, and Kolthoff and Nightingale[15] have shown that the degree of reversibility of the ferrous/ferric couple depends on whether or not the electrode has previously been oxidised. In the latter case, it is believed that the presence of an oxide film on a platinum electrode allows electron transfer between the electrode and the ions in solution more easily by means of an oxygen bridge formed between the oxide film and the hydrated iron ions.

From this information about the formation of oxide films on platinum electrodes, it is evident that care should be taken when using platinum or

References p. 165

gold electrodes in strongly oxidising solutions. Davis and Lingane[16] found that the potentiometric titration of the vanadyl ion with electro-generated Ag^{2+} was unsuccessful unless the indicator electrode had been previously oxidised. Generally, the formation of oxide films does not render titrations impossible, but rather causes equilibrium potentials to be established only slowly. Usually, any error can be avoided by allowing potentials to become stable before a reading is taken; this is not possible with some methods of automatic potentiometric titration in which it is assumed that equilibrium potentials are established rather rapidly. Serious errors could result, and any proposed automatic titration should therefore be tested manually to avoid such difficulties.

As well as being subject to oxidation, platinum electrodes are often found to be poor indicator electrodes in strongly reducing solutions[17]: for instance, it is generally better to use a mercury electrode for potentiometric titrations with chromous ions, because the chromous ion can reduce the hydrogen ion at a platinum surface; this causes the potential at the indicator electrode surface to be established by the hydrogen gas/hydrogen ion couple rather than by the chromous/chromic couple.

(b) *Apparatus*

(i) *Potentiometers*

The potential of a galvanic cell depends on concentrations of the chemical species involved in the electrode reactions as well as on their nature; these concentrations will change whenever electrode reactions occur and current passes through the cell. Because of this, a simple voltmeter is not useful for the exact measurement of potentials. A voltmeter allows an appreciable amount of current to flow, and this results in changing concentrations within the cell, and in the production of a potential drop that opposes the e.m.f. at the electrodes. It is evident then that, if potential measurements are to have any meaning, they ought to be made with a device that does not cause the passage of an appreciable current through the cell; such a device is a potentiometer that makes use of a compensation arrangement.

A typical potentiometer is illustrated in Fig. III, 6. A constant voltage from the battery (T) is impressed on the slide wire (AC), which has a uniform resistance, so that the voltage drop between A or C and any point B is proportional to the length AB or BC; the wire is provided with an accurate scale on which the position of the sliding contact (B) can be measured. The potentiometer is calibrated by bringing the standard cell (E_s) into the circuit by means of the double-pole double-throw switch (S). The sliding contact (B) is set at the scale value that corresponds to the known voltage of the standard cell. The potentiometer is balanced by alternately varying the resistance (R) and tapping the key (K) until the galvanometer (G) is no longer deflected; a key is used to avoid drawing any more current than necessary through the cell. After the potentiometer has been standardised, the unknown cell (E_x) is placed in the circuit by means of switch (S). The sliding contact (B) is again adjusted until, when the key is momentarily closed, no deflection of the galvanometer is observed.

This description is very much simplified in terms of both theory and practice. Generally, the maker's literature should be consulted for the exact operation of any particular potentiometer.

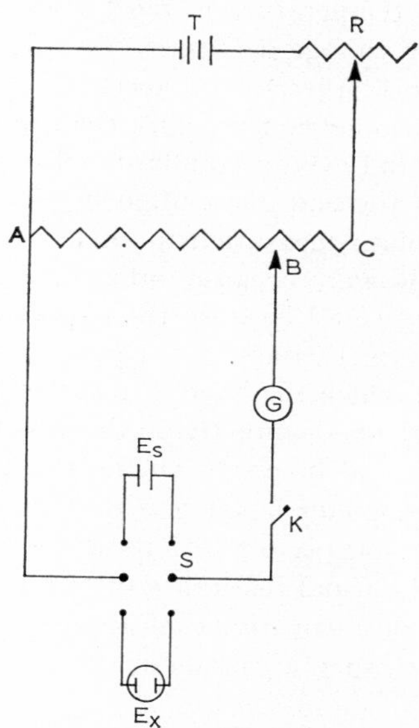

Fig. III, 6. Potentiometer.

The standard cell most normally used is the Weston cell, which may be represented by

$$(-)Cd(Hg)/3CdSO_4.8H_2O(sat.)\,;\; Hg_2SO_4(solid)/Hg(+).$$

The cell is usually assembled in a sealed H-shaped glass vessel, which is often contained in a wooden or plastic box. At any given temperature, the activities of all the various participants in the electrode reactions are constant, so that the voltage of the cell remains unchanged. If appreciable current is passed through a Weston cell, the activities undergo temporary changes that cause the voltage to change temporarily; the cell voltage will eventually return to its proper value after the re-establishment of the various

References p. 165

solubility equilibria involved. As this return to equilibrium may be quite slow, care should be taken to avoid the passage of large amounts of current through a standard cell.

Commercially available Weston cells are furnished with a standardisation certificate, so that the voltage is exactly known. These cells are usually made with a cadmium sulphate solution saturated at 4°, so that at normal temperatures the solution is, in fact, unsaturated. Calibration is made at some specific temperature recorded on the certificate, but as the temperature coefficient of such cells is about 10^{-5} volt per degree, the cells may be used over a range of temperature in most work. Even when Weston cells are well treated, their voltage decreases very slightly over a long period of time; for accurate work, therefore, a recently standardised cell should be used. For most potentiometric titrations, the simplest potentiometer arrangement is adequate, and even a standard cell is not strictly necessary, because the location of the end point often depends on changes in potential and not on the absolute measurement of it. Absolute measurements are sometimes made, however, because information about the reactions involved can be obtained and because the experimental results can then be compared with theoretical values.

The accuracy with which the balance point of a potentiometric measurement can be located depends directly on the sensitivity of the galvanometer (Fig. III, 6, G) used, and inversely on the resistance of the cell circuit. A simple pointer galvanometer has a sensitivity of about 10^{-7} A/mm, which allows a 1 mm deflection per 0.1 mv, if the cell circuit resistance is about 1000 ohms; this is the usual resistance of ordinary galvanic cells[18].

Galvanometers with a sensitivity of approximately 10^{-10} A/mm can be readily obtained; with such a galvanometer, a cell resistance of 1 megohm can be tolerated.

(*ii*) *Electronic Instruments*

When the sensitivities of galvanometers are too small for the purpose in hand (*i.e.* when the cell circuit resistance exceeds a megohm), an electronic instrument becomes necessary. Valve voltmeters have been developed in large numbers, especially for pH measurement with glass electrodes, the resistance of which sometimes approaches 100 megohms. Because these instruments draw only about 10^{-12} A from the cell, very high resistances are permissible. In addition, no tapping key is necessary, as the small amount of current drawn does not cause reactions at the electrodes to proceed to a noticeable extent (in most cases). Bates[19] has reviewed many of the various electronic circuits that have been developed, especially for the measurement of pH with glass electrodes.

Two general types of pH meters are available, direct-reading instruments and the "true potentiometer" type. In the latter, the electronic amplifier serves only as a null-point detector, whilst in the former, the amplifier changes the cell voltage to a signal that is indicated on a meter in pH units and millivolts.

Instruments of the true potentiometric type are exemplified by the familiar Beckman Model G and the very reliable Radiometer PHM4. In general, this type of instrument is more satisfactory for laboratory use, because it is more accurate and easier for a novice to repair. The fact that these instruments are battery-operated causes little trouble, since the batteries need replacement infrequently. In addition, because these meters operate independently, there is no chance of malfunction due to coupling of several instruments or appliances through the power line.

Despite these advantages, the potentiometric instruments are less popular than the direct-reading instruments because the latter are easier to operate; this is a distinct advantage, especially in potentiometric titrations when a number of readings must be recorded within a short time. The fact that such meters have become common laboratory equipment is undoubtedly responsible for the many applications of potentiometric titrations.

The simplest modern potentiometric titration apparatus consists of a titration beaker into which are lowered the electrodes of a direct-reading valve voltmeter. Stirring is most easily effected with a magnetic stirrer, and an ordinary burette serves to deliver the titrant. Most potentiometric titrations are facilitated by the fact that the meter makers generally supply a variety of indicator electrodes, such as glass, platinum, gold, silver and bismuth, and several reference electrodes as well.

A great variety of pH meters are manufactured, and it is thus impossible to mention all models or even makers. In the United States, the Beckman Instrument Company and Leeds and Northrup Company are especially noted; the latter also makes many fine potentiometers. One of the more recent instruments developed by Beckman is the Zeromatic pH meter shown in Fig. III, 7, which has not only a mirrored scale that is easy to read, but also an automatic zero-setting device, which gives it better stability than usual: every second, a corrector amplifier is made operative by the closing of a relay; the feedback part of the amplifier keeps the input grid of an electrometer valve at a constant potential while it furnishes energy to a storage capacitor. When the relay opens, the capacitor is charged to a voltage equal to the grid-to-earth potential of the electrometer valve, and the circuit is arranged so that this is the grid-to-earth potential of the pH measuring amplifier. The opening of the relay connects the capacitor to the grid circuit of the measuring amplifier in such a way that the signal input terminal is held

References p. 165

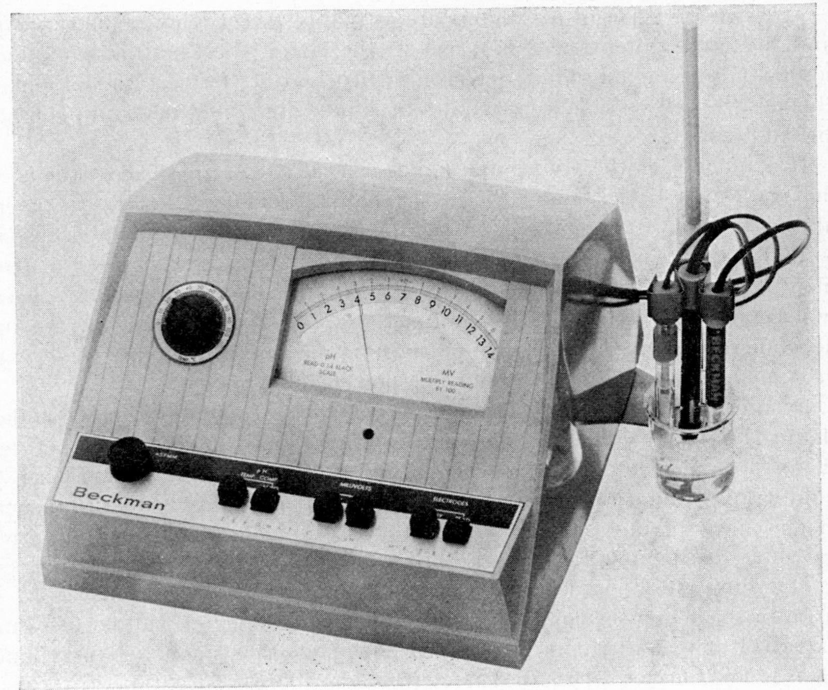

Fig. III, 7. Beckman Zeromatic pH meter.
(Courtesy of Beckman Instrument Company)

at earth zero. Thus, regardless of the bias drift of the input valve, the input is maintained at earth zero.

The Beckman Zeromatic also has a convenient recorder attachment, provision for the use of an automatic temperature compensator and a polarising circuit (polarising current about 10 microamp) that is specially useful for Karl Fischer titrations. The instrument can also be used for oxidation-reduction titrations, as can most pH meters. As a large variety of electrodes are available from the maker, the instrument is particularly suitable for all kinds of potentiometric titrations.

Beckman also makes a cation-sensitive electrode that is highly selective for univalent cations such as Na^+ and K^+.

The Radiometer pH Meter 22 (Radiometer, Copenhagen, Denmark) is also a meter of great convenience and accuracy. It is mains-operated, direct-reading, and has a self-adjusting zero. The large number of millivolt ranges available is an exceptional feature of this instrument, and makes it particularly useful for potentiometric titrations in which the change in potential is not great.

Another exceptionally accurate and stable mains-operated pH meter is the "Dynacap" manufactured by Pye and Company in Cambridge, England; this incorporates an expanded range device that allows the selection of various pH (or millivolt) ranges in such a way that a full scale deflection corresponds to only 2 pH units.

The meters listed above are not made particularly as potentiometric end-point detectors but, with few or no modifications, they are only exceeded in accuracy by fine potentiometers (which cannot be used with glass electrodes) and are almost unsurpassed in convenience for manual titrations.

Despite the fact that ordinary pH meters can readily be used for potentiometric titrations, a number of makers produce special manual titration apparatus designed primarily for control work: for instance, the Fisher Titrimeter (Fisher Scientific Company, Chicago, Illinois, U.S.A.) consists of a pH meter and a special stand that holds a magnetic stirrer and burettes; the electrodes connected to the pH meter are also held by the stand. To make the titration easier, the pH meter shows balance by means of a "magic eye" mounted on the titration stand. The pH meter can then be set at the end-point potential or pH (which must be calculated or found experimentally beforehand) and the addition of titrant is continued until the "magic eye" indicates that the solution potential, or the pH, is equal to that set on the meter. The titrimeter can be used for almost any potentiometric titration by proper selection of the electrodes. Titrations can be done automatically by the attachment of an inexpensive "titrating robot", which can be purchased as an accessory.

Fig. III, 8 shows the Gallenkamp Potentiometric Microtitration Apparatus[20], which is specifically designed for the semi-micro determination of chloride, but can undoubtedly be used for other work. The reference electrode is a piece of platinum wire sealed inside the burette, and the indicator electrode is the typical silver wire. The indicating device is a conventional double-triode valve voltmeter coupled with a short-period galvanometer.

Many other specialised titration devices are made. Among makers of particularly noteworthy apparatus in Europe are Metrohm, and J. Tacussel of Lyon, France, whose instrument is shown in Fig. III, 9.

For most purposes, a simple pH meter, a magnetic stirrer and a burette holder make the most versatile and inexpensive titration apparatus available. For special purposes, for example, where the number of titrations to be performed makes a pH meter too slow, or where exceptional accuracy, precision, or sensitivity are desired, other instruments may be of value.

References p. 165

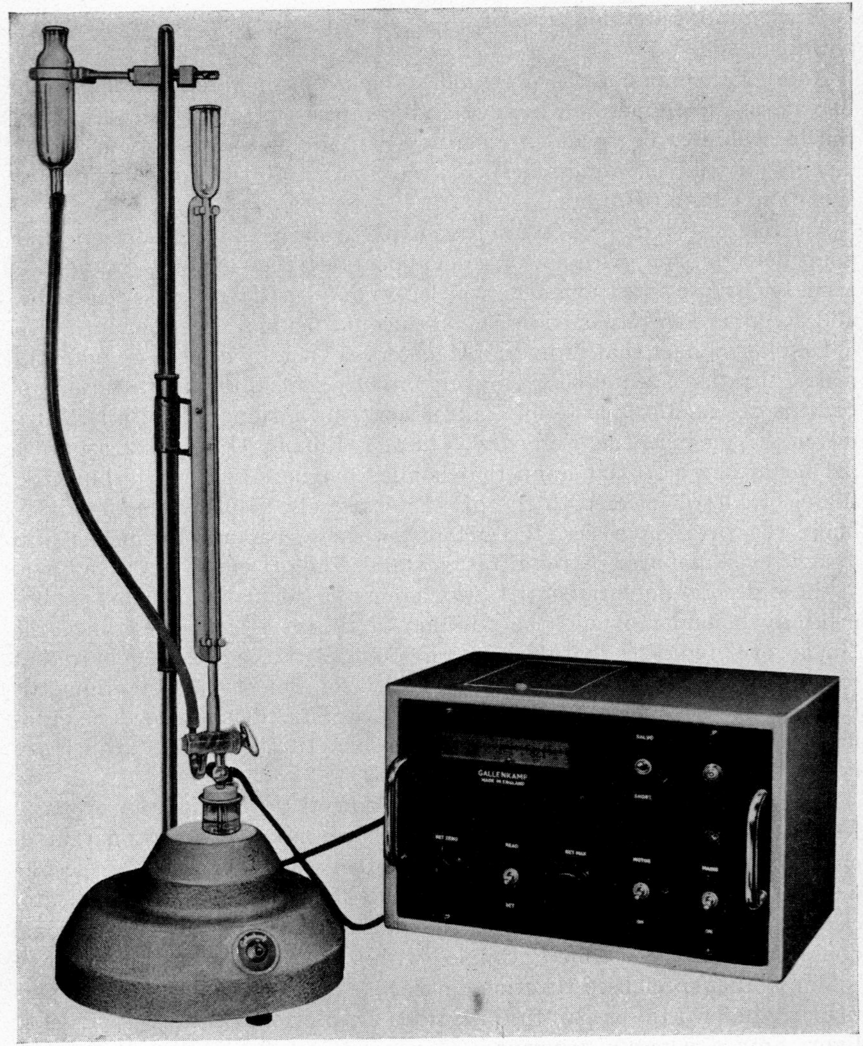

Fig. III, 8. Potentiometric microtitration apparatus.
(Courtesy of Gallenkamp Ltd.)

(*iii*) *Reference Electrodes*

Generally, potential measurements for analytical purposes are made with a cell consisting of a suitable indicator electrode immersed in the solution being investigated or titrated, and a reference electrode brought into contact

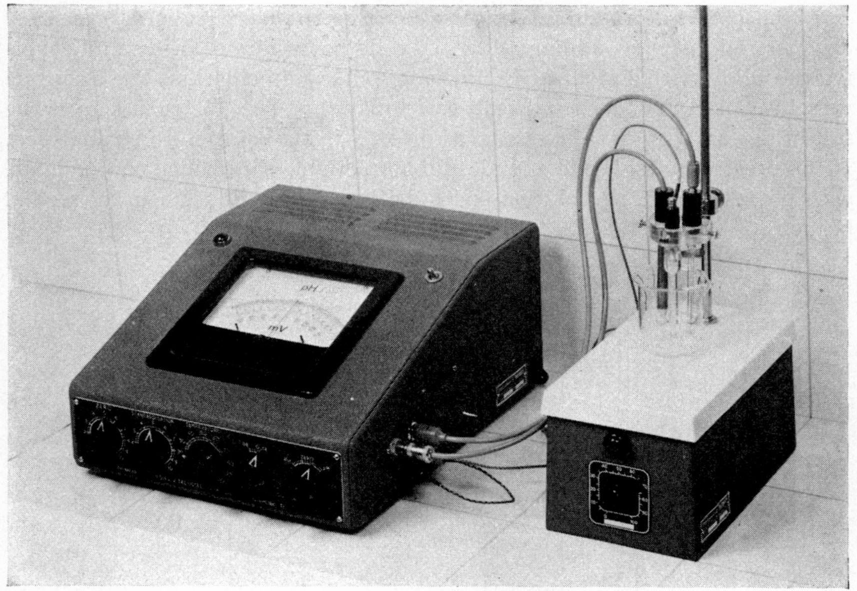

Fig. III, 9. Potentiometric titration apparatus.
(Courtesy of Société Lyonnaise d'Electronique Appliquée

with the solution by means of a "salt bridge". The hydrogen electrode, which is used to define the zero of the standard potential scale, is so inconvenient that it is not generally used in practical work. The most commonly used reference electrode is the calomel electrode, which consists of a mercury/mercurous chloride (calomel) electrode in a potassium chloride solution of some specified concentration. A saturated calomel electrode (containing saturated potassium chloride solution) is slightly easier to make and maintain than any other calomel electrode such as the N potassium chloride type. The potential of the saturated calomel electrode (S.C.E.) relative to the hydrogen electrode is

$$E = +0.2415 - 7.6 \times 10^{-4} (t-25) \text{ v},$$

where t is the temperature in degrees Centigrade.

Potassium chloride solution is an appropriate choice for the contents of a salt bridge, and the liquid-junction potentials developed are very small, because the main electrolyte in the reference electrode is potassium chloride and the mobilities of potassium ion and chloride ion are nearly equal.

Saturated calomel electrodes can be easily made by placing a layer of mercury in the appropriate electrode vessel, Fig. III,10, which is available

References p. 165

from most laboratory supply houses, covering the mercury with a paste of calomel and mercury, adding a layer of solid potassium chloride, and covering the whole with a saturated solution of potassium chloride. The electrode vessel is closed by a cork and sealed with paraffin. The "J" tube acts as the salt bridge and can usually be placed directly in the solution to be analysed. If the presence of chloride is undesirable, a double salt bridge can be made with potassium nitrate solution, instead of potassium chloride, in the second branch.

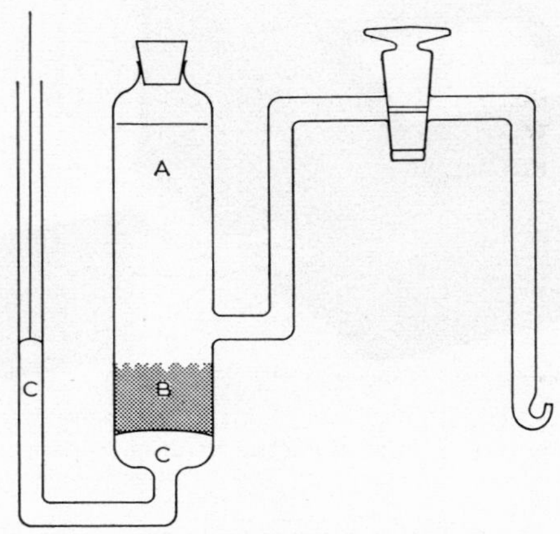

Fig. III, 10. Calomel Reference Electrode.
A, sat. KCl; B, calomel/mercury paste; C, mercury.

Silver/silver chloride electrodes, containing a saturated solution of potassium chloride and silver chloride in contact with metallic silver, are also used as reference electrodes. The potential of such an electrode is +0.197 v against the normal hydrogen electrode (N.H.E.). If chloride ion is to be avoided, a mercury/mercurous sulphate electrode is sometimes used; if it contains saturated potassium sulphate, its potential is +0.64 v against N.H.E.

Saturated calomel electrodes of a very convenient form are supplied with most valve voltmeters; these electrodes are relatively small and are placed inside a glass tube. Contact with the test solution is made by means of an asbestos fibre sealed in the bottom of the glass tube. As there is such a small area of contact between the reference electrode and the test solution, the solutions in each half-cell show little tendency to mix; there is some mixing, however, and even these electrodes should be avoided in such work as the titration of chloride with silver ion.

The resistance of these fibre type reference electrodes is rather high, but this is generally no problem when a pH meter is used as the measuring device. Nevertheless, electrodes of lower resistance, which make contact with the solution through a wet ground-glass joint, are available from many makers. These so-called "sleeve" type electrodes are particularly useful with solutions of high resistance such as are often encountered in non-aqueous titrimetry.

Quinhydrone electrodes (see p. 119) in buffers of known pH are sometimes useful reference electrodes, because any potential between about +0.7 v and +0.2 v *versus* the hydrogen electrode may be selected.

Because of the fact that end-point location is of primary concern in potentiometric titrations, it is often possible to use a glass electrode for a reference electrode in any precipitation, complex-formation, or redox titration during which the pH remains nearly constant. The glass electrode can only be used when the change in potential is all that is sought, unless the pH is exactly known and the meter is properly calibrated. It has been the author's experience that a glass electrode used as a reference electrode for halide titrations with silver often becomes fouled, and eventually increases in resistance so much that it can no longer be used.

(*iv*) *Automatic Potentiometric Titrations*

Because titration is such a common analytical operation, extensive work has been devoted to making it automatic, and potentiometric titrations have been particularly studied. In general, automatic potentiometric instruments can be divided into three classes: in the first, the titration curve is simply recorded automatically and the end point is assumed to be the inflection point of this curve; in the second method, some provision is made for stopping the flow of titrant when the indicator electrode potential reaches a certain pre-selected value; the third method employs an electronic circuit that essentially differentiates the titration curve and stops the flow of titrant when a large change in dE/dV or d^2E/dV^2 occurs.

The method of recording the complete titration curve and reading the end point from the recorded curve has certain disadvantages. Because of mixing delays and the fact that a measurable amount of time is necessary for many chemical reactions to reach equilibrium, the accuracy of this method is not great; in fact, slow reactions cannot be used without much error. Robinson[21] designed an instrument that minimises this source of error: the recorder controls the delivery of small increments of titrant in such a way that equilibrium is attained before the addition of each succeeding increment; an instrument similar to Robinson's is available from the Precision Scientific Co., Chicago, Illinois, U.S.A. as the Precision-Dow Recordomatic Titrator.

The method of titrating to a set value of the indicator potential was first used by Ziegal[22]. The instrument devised by Shenk and Fenwick[23] has an electronic detector that operates a solenoid to stop the flow of titrant when the indicator electrode potential reaches the equivalence-point value. The best results are obtained if reagent is added very slowly, but errors inevitably result, just as they would if a titration were attempted with the stopcock of the burette open up to the end point.

Lingane[24, 25] realised that the accurate automatic detection of a potentiometric end point depends on having some means of anticipating the end point and repeatedly adding small increments of titrant until the equivalence-point equilibrium has been reached throughout the entire solution. Lingane's autotitrator has a motor-driven syringe to deliver the titrant, and the capillary delivery tip is placed under the surface of the solution close to the indicator electrode. The anticipatory approach is provided in this way, because the indicator electrode is in contact with a solution that is at a more advanced stage of titration than the bulk of the liquid; the distance between the indicator electrode and the delivery tip is chosen according to the rate of the titration reaction.

The potential difference between the indicator and the reference electrodes is presented to an electrical device that controls the motor driving the syringe. Near the end point, the motor circuit is opened when the indicator electrode potential reaches a value close to the equivalence-point potential. When mixing is complete and the potential falls to a certain value, the delivery of titrant starts again. In this way, increments of titrant are added until the equivalence point is reached.

The Beckman Auto-titrator Model K (Fig. III,11) has been completely described by Hawes, Stricker and Peterson[26]. The measurement of the indicator electrode potential (*versus* the reference electrode) is made by means of a null-sensing amplifier that powers a relay controlling the electromagnetically actuated burette valve; when the indicator electrode potential reaches the equivalence-point potential, the relay closes the burette valve. An incremental approach to the equivalence point is effected by a controllable anticipation circuit. Proper adjustment of the anticipation circuit combined with proper placement of the burette tip allows titrations with an accuracy limited only by the accuracy with which the burette can be read. This instrument is particularly good for the titration of a large number of similar samples; to this end, the instrument is supplied with four titration vessels, electrode pairs, *etc.*, so that four titrations can be done in rapid succession.

The automatic titration device manufactured by Radiometer (Fig. III, 12) is applicable to both research and routine work. The unit shown is actually a combination of a "Titrigraph" and a "Titrator TTT 1". This combination is particularly versatile, in that titration curves of all kinds may be recorded

Fig. III, 11. Beckman Autotitrator.

(Courtesy of Beckman Instrument Company)

automatically. For routine work, the titrator alone is sufficient. The method of operation is basically similar to that of the Beckman Autotitrator, but the Radiometer instrument can be set to the end-point potential with somewhat greater accuracy; in addition, the magnetic valve is designed to control the flow from any laboratory burette. At rest, the valve throttles the valve tubing but, when it is actuated, it allows the titrant to flow at a pre-adjustable rate. The instrument is supplied with an automatic shut-off, which turns off the control stage after a pre-selected time has elapsed without further addition of titrant. One of the settings on the instrument is that of "proportional band": in the proportional band, the average flow of titrant is proportional to the difference between the end-point setting and the actual reading; the drop-by-drop approach to the end point is thus performed automatically.

The combination of the Titrigraph and the TTT 1 is provided with an all-glass syringe burette. The deflection of the recorder follows the setting of the piston of the burette, and is thus proportional to the volume of titrant

References p. 165

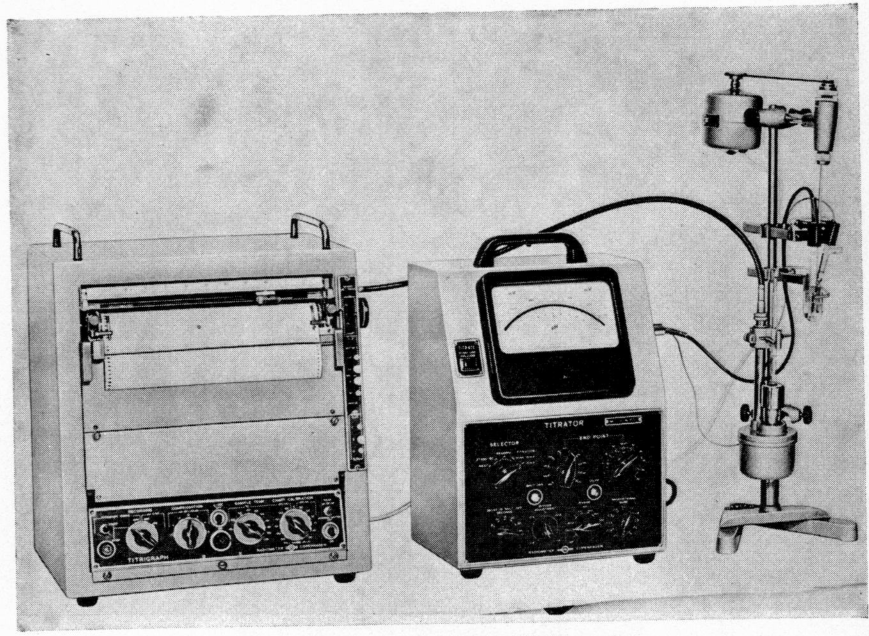

Fig. III, 12. Titrator and Titrigraph.

(Courtesy of Radiometer Inc.)

added; the paper drive follows the variation of the indicator potential. Titration curves can be made very rapidly, because the rate of addition of titrant adapts itself to the slope of the titration curve. The instrument can also be set to record the consumption of titrant at a constant pH or potential, and thus serves the function of a very stable "pH-stat".

The Pye Automatic Titrator (W. G. Pye and Co., Ltd., Cambridge, England), shown in Fig. III, 13, also titrates to a pre-determined equivalence-point potential or pH. The burette of this instrument is provided with fast and slow delivery jets, and during the first part of the titration both are open so that the titration proceeds rapidly. At a pre-determined potential, the fast jet is closed and the titration continues by dropwise addition from the slow jet; the range of this anticipation control is 5 pH units or 500 millivolts.

Differential titrators depend upon observations of the slope of the titration curve, and are generally designed to stop the flow of titrant when the inflection in the titration curve is reached. Baker and Müller[27] carried out differential titrations with an electronic pulse amplifier; the reagent was added continuously at a constant rate and the end point was indicated by a sharp maximum pulse.

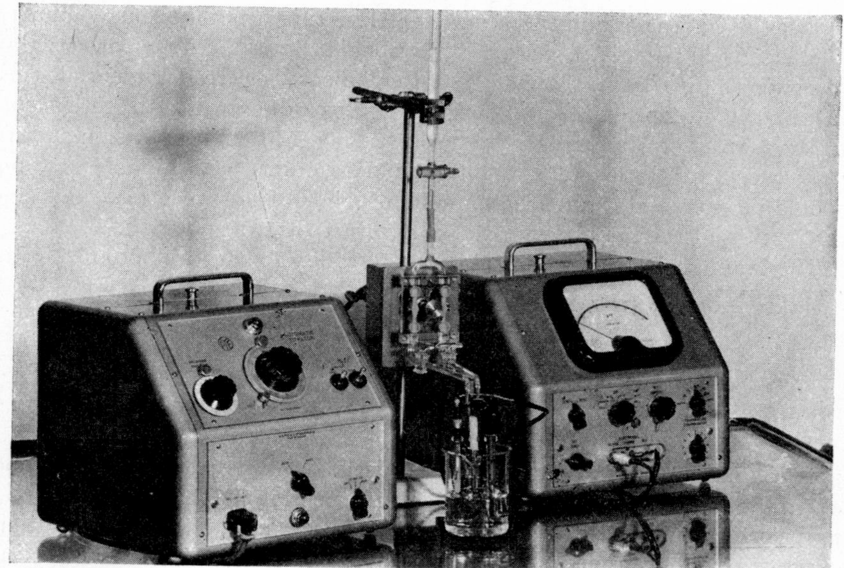

Fig. III, 13. Pye Autotitrator.

(Courtesy of W. G. Pye Co., Ltd.)

Malmstadt[28] and his students have developed an automatic titrator based on the principle of the capacitive coupled amplifier. This instrument responds to the second derivative, d^2E/dV^2, of the titration curve by the addition of a second amplifier and differentiator to the output of the first derivative stage; the result is that the time-varying input signal is differentiated twice. As the second derivative of the titration curve changes sign at the end-point, the output of the amplifier does the same. The second derivative voltage is applied to the grid of a thyratron valve that forms one element of a relaxation oscillator. When the derivative signal rises above a certain value, the oscillator begins to oscillate at constant amplitude until the voltage falls to the same level. A filter circuit removes the halfwave oscillations, thus producing a square wave, which is differentiated by a circuit similar to those used for the first differentiations; the effect of this is to produce one positive and one negative pulse. The positive pulse is applied to a second thyratron valve, which acts as a switch to shut off the burette[29]. The Malmstadt titrator is manufactured by E. H. Sargent Co., Chicago, Illinois, U.S.A.

The advantage of automatic differential titrations is that the equivalence-point potential need not be known, and this greatly simplifies the design and the operation of the instrument. It can be used with electrodes whose behav-

References p. 165

iour is too variable for direct potentiometric use, such as, for example, bimetallic electrode pairs[30]. It must be remembered that the instrument actually responds to the *time rate* of change of the indicator electrode potential; this rate of change depends on such factors as the rate of titration, the rate of the titration reaction, and the conditions of stirring. Although the instrument is very successful with a great many rapid reactions, considerable caution should be exercised in the location of end points of slow reactions; this is a distinct disadvantage as compared with direct titrators.

The Malmstadt titrator has also been applied to spectrophotometric titrations[30, 31].

(c) *End-Point Detection of a Potentiometric Titration*

(i) *Graphical Determination*

The easiest and most obvious way of locating the equivalence point of a potentiometric titration is to plot the potential or pH as a function of volume of titrant added, and then to determine by inspection the value of the volume that corresponds to the maximum slope of the curve. When the titration curve is perfectly symmetrical, the maximum value of $\mathrm{d}E/\mathrm{d}V$ is identical with the stoichiometric equivalence point. A symmetrical titration curve will be obtained when the indicator reaction is reversible and the titration reaction is symmetrical. If the titration reaction is not symmetrical, the maximum value of $\mathrm{d}E/\mathrm{d}V$ will not occur at the exact equivalence point. The titration error is usually small, provided that there is a reasonably large potential change near the equivalence point, and it may be calculated exactly, provided that the indicator electrode reaction is reversible[32].

The end point can be rapidly and exactly located by differentiation rather than by actually plotting a graph. The end point is found by adding small equal increments of titrant near the end of the titration and recording the indicator electrode potential; the maximum value of $\mathrm{d}E/\mathrm{d}V$ is then computed by taking it as the point where $\mathrm{d}^2E/\mathrm{d}V^2$ becomes zero[33]. A typical example is given on p. 103.

Fig. III,14 illustrates the principle underlying this method. The best value of $\mathrm{d}V$ depends on the slope of the titration curve near the equivalence point. In general, the greater the slope at the equivalence point, the smaller the value of $\mathrm{d}V$ must be. It must be admitted, however, that practical considerations govern the choice of $\mathrm{d}V$ as much as theoretical ones. The value of $\mathrm{d}V$ must be large enough for successive values of $\mathrm{d}E$ to show a significant difference. In an ordinary titration, it is troublesome to make $\mathrm{d}V$ smaller than 0.05 ml ($\sim$ 1 drop). If a particularly sharp titration curve is encountered, this may be still too large for the application of this method; this is of no

Volume	E	$\mathrm{d}E/\mathrm{d}V$	$\mathrm{d}^2E/\mathrm{d}V^2$
12.00 ml	0.420 volt		
		5 mv/0.1 ml	
12.10	0.425		+ 7
		12	
12.20	0.437		+20
		32	
12.30	0.469		+74
		106	
12.40	0.575		−36
		70	
12.50	0.645		−20
		50	
12.60	0.695		

$$V = 12.30 + 0.10 \left(\frac{74}{74 + 36}\right) = 12.37 \text{ ml}$$

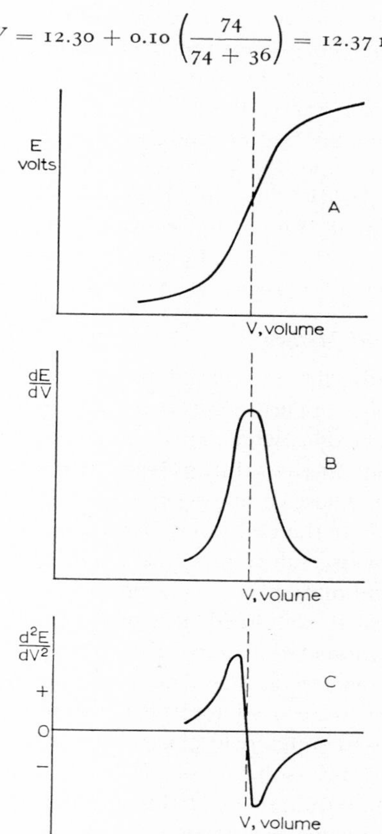

Fig. III, 14. Titration curve and 1st and 2nd derivative curves.

References p. 165

practical consequence, however, because in such a case the end point is approached dropwise, and the volume reading at the largest potential change is taken as the stoichiometric amount of titrant.

(ii) Titration to the Equivalence-Point Potential

When the potential of the indicator electrode at the equivalence point is known beforehand, or can be calculated, the titration can easily be carried out by adding the titrant until this potential is reached[34]; this method is analogous to an ordinary titration carried out to a specific indicator colour. The potentiometer is simply set to the value of the equivalence-point potential and the titrant is added until the galvanometer shows no deflection when the key is momentarily closed. The method is extremely rapid and convenient, because it is not necessary to record results or make calculations or graphs. If the equivalence-point potential cannot be calculated, it can be found experimentally by carrying out a titration and plotting a curve from the values obtained near the end point.

The accuracy of this method depends on the reproducibility of the equivalence-point potential, and the relative accuracy with which it is known in comparison with dE/dV. The larger dE/dV is, the less accurately the equivalence-point potential must be known. Cooke, Reilley and Furman[35] have found that this method of end-point detection is especially useful for titrations involving small amounts of material.

(iii) Pinkhof-Treadwell Method

The Pinkhof-Treadwell[36, 37] method of end-point detection is actually a variation of the equivalence-point potential method. The titration cell is made up of two electrodes, which are often identical; one of these serves as the indicator electrode in the solution being titrated, and the other, which is placed in a solution whose composition is the same as that which the titrated solution will have at the end point, serves as a reference electrode. Contact is made between two solutions by means of an appropriate salt bridge. As this electrode combination has a potential of zero at the end point, the apparatus can be greatly simplified: it is only necessary to connect the two electrodes to a galvanometer in series with a resistance and a tapping key and titrate until no deflection of the galvanometer occurs. If, for example, a hydrogen electrode were used for the titration of a strong acid with a strong base, the second (reference) electrode would consist of a second hydrogen electrode in a buffer solution of pH 7.

Actually, the basic requirement of this method is that the reference electrode should have the same potential as the indicator electrode at the end point, and the reference electrode need not, therefore, be of the same type as the indicator electrode. Chloride ion can be titrated with silver ion by using a silver electrode and a quinhydrone reference electrode; the potential at the end point of the silver electrode should be +0.270 V *vs.* S.C.E. The quinhydrone electrode should be placed in a buffer of pH 3.1 to ensure that the cell potential is zero at the end point. The potential of a quinhydrone electrode *vs.* S.C.E. is given by: $E = 0.453 - 0.0591$ pH, at 25°.

For normal oxidation-reduction reactions, two identical platinum electrodes would be used. The indicator electrode would of course be placed in a solution being titrated, while the reference electrode would be placed in a solution that is free from the ion to be determined, but otherwise identical in composition with the test solution (or, sometimes in a solution of the titration products) and its potential brought close to that of the end-point potential by the addition of one drop of titrant. Thus, for the titration of ferrous iron and vanadyl solutions, Lang[38] recommends a reference electrode in contact with a solution containing one drop of $0.1N$ potassium permanganate in 50 ml of $1\text{-}3M$ hydrochloric acid.

While the method of Pinkhof and Treadwell is very rapid and entails simple apparatus, care must be exercised in its use when, for any reason, there is the possibility that the potential may be only slowly established. As the potential of the indicator electrode is not measured, the operator knows nothing about the titration, except whether or not the end point has been reached. This method should generally be used only for the routine performance of titrations that are known to be characterised by the relatively rapid establishment of constant potentials.

(*iv*) *Differential Titrations*

Differential titration is based on the direct measurement of dE/dV as a function of the volume (V) of titrant added. Cox[39] suggested that two identical portions of the solution to be analysed should be placed in two beakers connected by a salt bridge, and the potential difference between identical indicator electrodes placed in each beaker measured. Each solution is titrated with the same standard solution delivered from its own burette, and the titration is carried out in such a way that one of the burettes has delivered 0.1 or 0.2 ml more than the other. The end point is indicated by a sharp rise in the difference in potential between the two indicator electrodes (Fig. III, 14B). Although this procedure is somewhat complex, its advantages are that no curve need be plotted and no calomel or other reference electrode is needed.

MacInnes and his co-workers[40–42] have developed differential titrations to such an extent that they are now quite simple and practical (for an extensive discussion see reference 43). Two identical electrodes are used, one

References p. 165

being isolated in a compartment containing only a small portion of the solution being titrated. A simplified version of such apparatus is shown in Fig. III,15. The isolated or "retarded" electrode is sealed inside a small medicine dropper provided with a rubber bulb for flushing and refilling the isolation compartment. The titration is carried out by adding an increment of titrant and measuring the value of dE between the electrodes. The rubber bulb is used to flush the isolated electrode compartment and the

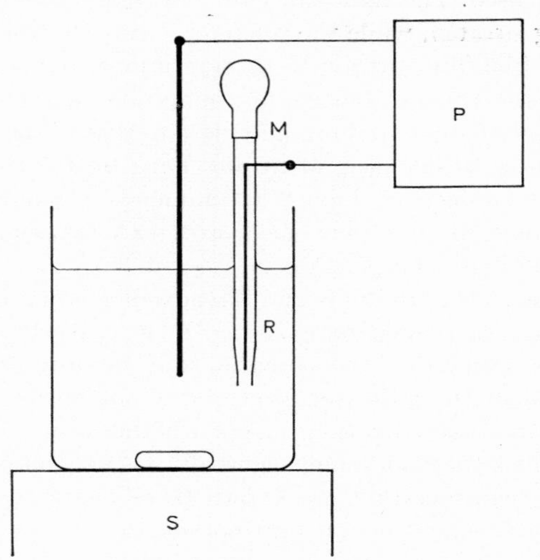

Fig. III, 15. Differential titration apparatus.

next increment of titrant is then added. In this way, the solution in the isolated compartment is at a stage of the titration slightly behind the main solution. The volume of the isolated compartment is kept small, so that the withholding of this volume of solution causes a negligible error in the amount of titrant used.

Instead of a potentiometer, a simple galvanometer connected to a high resistance and a tapping key can be used to measure the difference in the potential of the electrodes, because absolute potential values are not needed and, near the end of the titration, the value of dE increases gradually enough to give warning of the approach of the end point.

The main advantage of this method is that no reference electrode is needed. Its accuracy is as good as, but no better than, that of other methods of potentiometric end-point detection. Actually, the convenience and speed of this method is less than that of titrating to the equivalence-point potential.

(v) Bimetallic Electrodes

Hostetter and Roberts[33] observed that a palladium electrode, unlike a platinum electrode, does not respond to changes in potential in the ferrous-dichromate titration. They found that the potential difference between a platinum and a palladium electrode in a ferrous solution being titrated with dichromate is small until the end point is reached; at the end point, a large increase in potential difference is observed. Thus, the palladium electrode may replace the calomel or other reference electrode. An even larger potential difference is observed with the platinum/tungsten couple.

The difference in potential between the platinum and palladium electrodes appears to be due primarily to the fact that the rate of potential establishment is much slower at the palladium electrode. The observed increase in potential at the end point varies with time, and titrations with bimetallic electrode systems are therefore best done rapidly.

Kolthoff[44] has pointed out that the tungsten electrode behaves as an "attackable" reference electrode, in that a film (WO_3?) is formed on a tungsten electrode by the passage of anodic current. Generally, the extent of the attack of the tungsten electrode will not be large but, for this and other reasons, the tungsten/platinum couple is not generally suitable for use with dilute solutions.

Many combinations of metal electrodes have been studied[45–54]. Most often, one of the metals is platinum (indicator) and the second metal is either slow to respond to potential changes or is "attackable". Essentially, all types of reactions have been considered.

The use of bimetallic combinations is empirical, and optimum conditions must be established by trial and error. In addition, the pre-treatment of electrodes is generally very important. For these reasons, the use of such systems has little advantage over the use of the usual indicator/reference electrode pair, except possibly in those situations, such as some non-aqueous titrations, where a reliable reference electrode cannot be found.

(vi) Potentiometry with Polarised Indicator Electrodes

A number of the generally used oxidation-reduction couples, such as $Cr_2O_7^{2-}/Cr^{3+}$, are slow in establishing constant potentials at a platinum indicator electrode, when measurement is made with an ordinary balanced potentiometric circuit. It is often possible to avoid long waiting periods by forcing slight electrolysis to occur at the indicator electrode. A large number of studies of the potential changes of polarised electrodes with various oxidation-reduction titrations have been carried out[45, 46, 54–60]. Kolthoff[44] has discussed the interpretation of potentiometric titrations with polarised electrodes in relation to current-potential curves, and has divided the methods

References p. 165

used into two classes, depending on whether one or two indicator electrodes are used. In the former, the potential of a single polarised platinum indicator electrode against a reference electrode is measured; this indicator electrode may be polarised anodically or cathodically according to the situation. If two platinum electrodes are used, one will act as a cathode and the other as an anode, and the potential between the two is measured; Bishop[58] has called this method "differential electrolytic potentiometry."

If both the couples involved in the titration are reversible, the potential change at the end point of one polarised indicator electrode will be about the same as when the electrode is not polarised. When, however, one or both of the couples involved are irreversible, the potential change may be much greater if the indicator electrode is polarised.

Despite the advantages of using one polarised electrode when electrode potentials are established slowly, the experimenter must use this method with caution, because the point of greatest potential change will not correspond to the equivalence point, with the result that an error that is proportional to the size of the polarising current may be incurred[55]. In addition, the composition of the solution may be altered by electrolysis but, as currents of the order of only a few microamps are generally used, this error will be negligible at ordinary concentrations. The use of two polarised electrodes has the same disadvantage, if one of the couples involved in the titration is irreversible. If both couples are reversible, the generation of oxidant at the anode will be just balanced by generation of reductant at the cathode. The use of polarised electrodes has, however, no advantage if both couples are reversible, since steady potentials will be rapidly established at an unpolarised electrode.

The interpretation of titration curves obtained with polarised electrodes generally depends on the study of current-potential curves of the oxidation-reduction couples involved; a simple example is the titration of thiosulphate with iodine: the thiosulphate/tetrathionate couple is irreversible, while the iodine/iodide couple is reversible. Fig. III, 16 shows the current-potential curves that would be measured at different stages of this titration. Curve 1 represents all stages of the titration up to and including the equivalence point; being irreversible, the thiosulphate/tetrathionate couple will not react at a platinum electrode; thus, before the equivalence point, an anode would oxidise the iodide ion, whilst a cathode would reduce oxygen (or if the solution were air-free, hydrogen ion or water). An unpolarised electrode would establish a potential somewhere on the horizontal portion of curve 1. The potential of the unpolarised electrode would be somewhat indeterminate, as no reversible couple would be present.

After the equivalence point, an excess of iodine would be present, and the situation would be expressed by current-potential curves 2, 3, or 4, depending on the excess added; curves 2 and 3 show a levelling off at specific values of cathodic current, and these values are diffusion currents, I_d, indicating that the amount of iodine being reduced is limited by diffusion. The exact value of I_d will depend, not only on the concentration of iodine, but also on the electrode area, rate of stirring, temperature, *etc.*, and these variables must be considered in setting up a potentiometric titration with polarised electrodes.

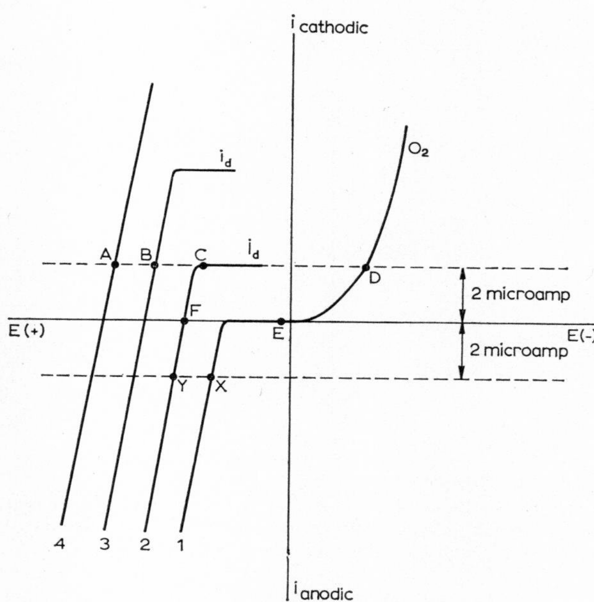

Fig. III, 16. Current-potential curves of the iodide-iodine couple. (1) No iodine, large concentration of iodide, (2) $10^{-5}N$ iodine, large concentration of iodide, (3) $10^{-4}N$ iodine, large concentration of iodide, (4) Large concentration of iodine, large concentration of iodide.

If one unpolarised indicator electrode is used, its potential will jump from E to F (Fig. III, 16) and a curve such as curve 1, Fig. III, 17 will be obtained. The drawbacks here are that the distance between E and F is not great, and that the value of E is not well established; the curve will thus show only a small break, and steady potentials will not be established quickly before the equivalence point.

A single indicator electrode can be polarised in two possible ways, cathodically or anodically. If the single electrode is made an anode, a very small

References p. 165

break at the equivalence point will be observed (Fig. III, 17, curve 2), and the reason for this may be seen from Fig. III, 16: if the indicator electrode is forced to allow an anodic current of 2 microamp to pass, the potential before the equivalence point is reached is indicated by X; when an excess of iodine is added, only a slight jump to Y will occur.

From an analytical point of view, cathodic polarisation is much more useful, because the end point is indicated by a large jump in potential as

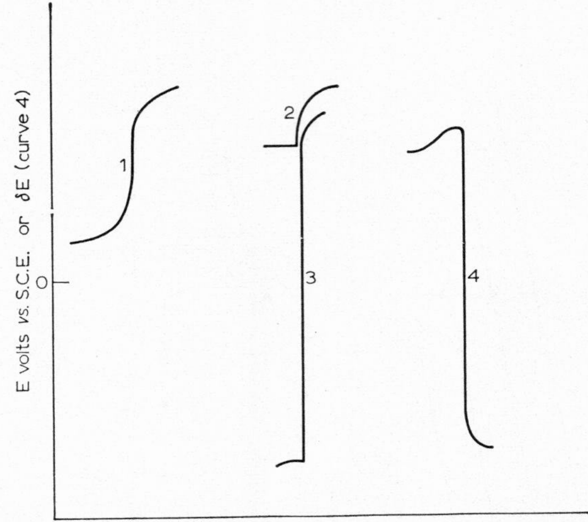

Fig. III, 17. Titration of thiosulphate with iodine. (1) One unpolarised indicator electrode, (2) One polarised indicator electrode (anode), (3) One polarised indicator electrode (cathode), (4) Two polarised indicator electrodes.

shown in Fig. III, 17, curve 3. Before the equivalence point, the indicator electrode will be acting as a cathode and reducing oxygen (point D in Fig. III, 16), but as soon as an excess of iodine is present, the potential will suddenly jump to C, where the indicating cathode can reduce iodine. It should be noticed that the jump in potential occurs only when an excess of iodine is present, and the end point will thus be slightly after the equivalence point; this causes only a slight error, if the polarising current is not too large.

If two identical platinum indicator electrodes are polarised (one being a cathode and the other an anode), a curve such as 4 in Fig. III, 17 will be obtained for the thiosulphate/iodine titration, and the shape of this curve may be interpreted by again examining Fig. III, 16: before the equivalence point, the indicator cathode will have a potential indicated by D, while the indicator anode will have a potential X, provided that the polarising current is about two microamperes; the potential difference will be even greater, if a larger current is passed, but it is unwise to use too large a polarising current, because the electrolytic generation of titrant may cause an error. When an excess of iodine (Curve 2, Fig. III, 16) is present, the potentials of the two indicator electrodes will suddenly draw very close to each other, the cathode at C and the anode at Y.

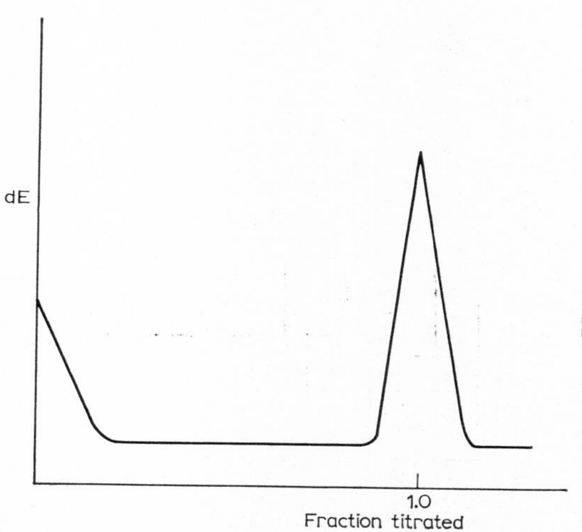

Fig. III, 18. Titration curve with two identical indicator electrodes polarised at constant current. (Both titrant and titrate couples reversible.)

These titration curves shown in Fig. III, 17 are fairly typical of an irreversible couple being titrated with a reversible. If a reversible couple is titrated with an irreversible, other things being equal, the curves will be reversed. It is evident that the important part played by the electrochemistry of each species must be considered if titration curves are to be predicted.

If both couples are reversible, one polarised electrode (either anode or cathode) will behave very much as an unpolarised electrode. If two polarised electrodes are used for the titration of one reversible couple with another, a curve such as Fig III, 18 is obtained. If, for example, ferrous ion is titrated with ceric ion in strong sulphuric acid, there is only a slight difference between the electrode potentials at the start of the titration and at the equivalence point, because the indicator cathode is reducing ferric ion, whilst the indicator anode is oxidising ferrous ion. At the equivalence point, no ferrous ion remains, as it has been oxidized to ferric by the ceric ion;

References p. 165

thus, the potential of the indicator anode will move so that it can oxidise some other substance, in this case cerous ion. After the equivalence point, the cathode potential will move to a value close to that of the anode, because the cathode will now have an excess of ceric ion to reduce, and a peak of potential difference will thus occur at the equivalence point; the size of this peak will be roughly the same as the difference between the formal potentials of the couples involved.

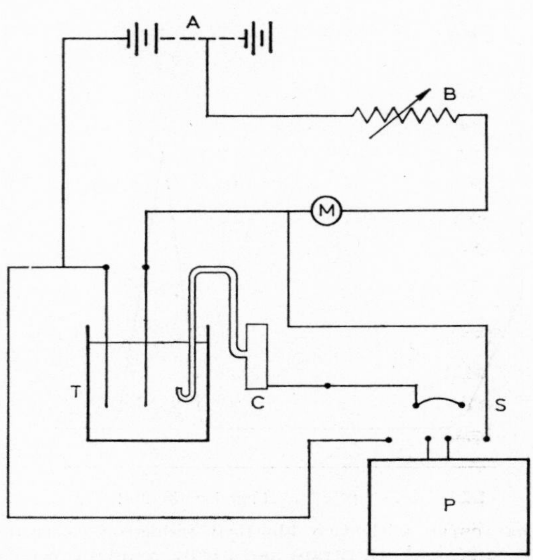

Fig. III, 19. Circuit diagram for potentiometric titrations with polarised electrodes.

Apparatus for titrations with polarised electrodes consists simply of a potential measuring device, such as a pH meter, and a circuit to pass a small constant current between the electrodes. Many pH meters now have a circuit built into them to provide such a small current, though only one current value is usually available. A very versatile arrangement has been proposed by Bishop[59], a simplified version of which is shown in Fig. III, 19; it is important to choose a battery combining a high voltage with a large series resistance in order to keep the polarising current essentially constant.

5. Applications

During the first sixty years of the century, about three thousand papers were published on the use of potentiometric end-point detection. In a work of this size it is impossible to consider all the methods developed, but an attempt will be made to discuss, or at least to mention, as many of the interesting and useful procedures as possible. The monographs by Müller[34], and by Kolthoff and Furman[61] contain much information and many practical procedures and, together with the reviews of Furman[62–65], and Reilley[66, 67], adequately cover the subject up to 1958.

(a) Acid-Base Titrations

(i) The Hydrogen Electrode

The hydrogen electrode was the earliest to be used, and it is the standard pH electrode. It consists of a noble metal electrode dipping into a solution saturated with hydrogen gas. One of the most elegant hydrogen electrodes was designed by Wood *et al.*[68], but the design of Hildebrand[69], shown in Fig. III, 20, is often used because of its convenience and simplicity. The holes (H) in the bell allow the liquid to cover about one half of the platinised platinum electrode (P). Hydrogen is allowed to pass continually through the electrode chamber. As the actual partial pressure of H_2 is not well defined, the electrode is suitable only for end-point location, and not for absolute measurement of pH. Kolthoff[70] found that a pointed platinum wire reached equilibrium more rapidly, and Cornish[71] increased the rate at which the potential was established by spraying the solution on to the electrode by means of an atomiser. Treadwell and Weiss[72] made a hydrogen electrode from a porcelain tube soaked in gold chloride, heated to reduce the gold, and finally plated with palladium black; two of these electrodes were used for a titration, one being placed in a buffer solution of a pH corresponding to the end-point pH (Pinkhof-Treadwell method). Palladium electrodes frequently prove to be unreliable, and platinised platinum is generally recommended[73].

Schwing and Rogers[74] have, however, studied various forms of the palladium electrode, and have found the most satisfactory to be one on which hydrogen was continuously generated. Stock, Purdy and Williams[75] designed a useful form of the "generation" electrode, and found that with "generation" currents of about 1 mA/cm^2, the relationship between pH and e.m.f. is nearly linear; they have also applied the palladium-hydrogen "generation" electrode to the titration of a number of acids in aqueous solution and in several solvents.

An electrode of the Kolthoff type is probably the best for general use, even though a cell such as shown in Fig. III, 21 must be constructed. It is not absolutely necessary for hydrogen to pass over the electrode all the time, provided that the solution is saturated with hydrogen. The continuous

References p. 165

bubbling of hydrogen does, however, provide a convenient means of stirring the solution during a titration.

Equilibrium of the electrode reaction

$$2H^+ + 2e = H_2 \tag{49}$$

is established rather slowly at a bright platinum surface, and platinum electrodes are generally coated with a finely divided layer of electrolytically

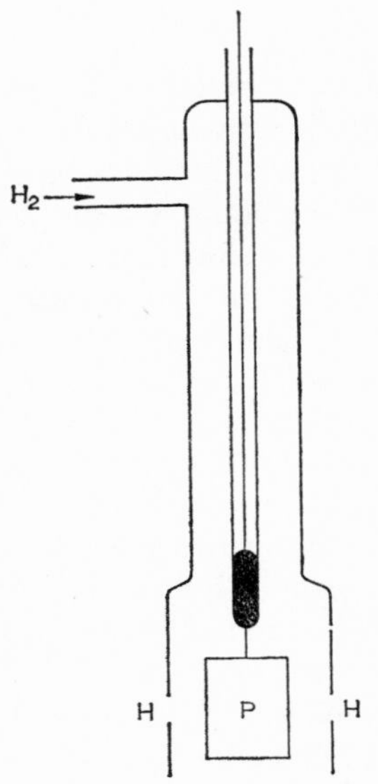

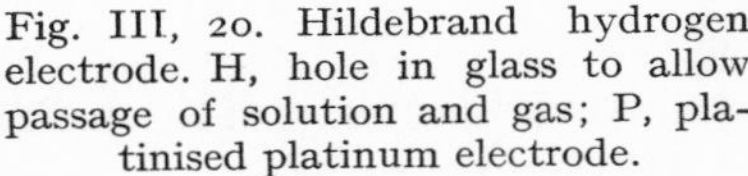
Fig. III, 20. Hildebrand hydrogen electrode. H, hole in glass to allow passage of solution and gas; P, platinised platinum electrode.

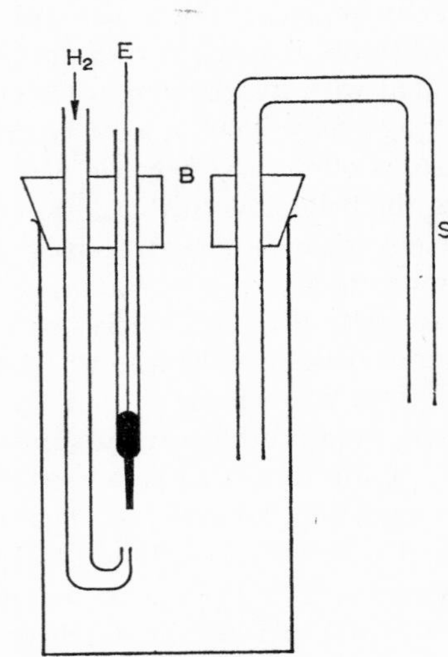

Fig. III, 21. Titration cell for use with a hydrogen electrode (Kolthoff). E, electrode; B, hole for insertion of burette and escape of hydrogen; S, salt bridge.

deposited platinum black. The effective area is also increased by this process, thus minimising changes in hydrogen ion concentration that occur at the electrode surface when current is drawn in the process of potentiometric measurement. The platinising process is effected by making a bright platinum electrode a cathode in a solution of chloroplatinic acid containing 1% of platinum until the electrode becomes dark grey. The coating of platinum black should be rather thin, because thick black coatings tend to adsorb cations, with the result that irregular readings are obtained, especially in slightly buffered solutions[73].

After being platinised, the electrode should be polarised cathodically in dilute sulphuric acid, carefully washed to remove impurities such as chloride, and stored in distilled water. When the electrode attains equilibrium slowly, the platinum black coating should be replaced.

The pH of the solution and the observed potential are related by the expression:

$$\mathrm{pH} = \frac{E_{\mathrm{obs}} - E}{0.05915} - \frac{1}{2}\log_{10} P_{\mathrm{H_2}} \tag{50}$$

at 25°, where E is the reference electrode potential plus any liquid junction potentials involved. In most cases, the value of $P_{\mathrm{H_2}}$, which should be expressed in atmospheres, may be calculated from the atmospheric pressure and the partial pressure of water vapour under the conditions of the experiment. For very accurate work, the effect of solutes in the water and the depth to which the bubbling tube extends must be considered.

Traces of oxygen have a great effect on the potential of a hydrogen electrode; care must therefore be taken to exclude oxygen from the solution by the passage of hydrogen for several minutes before a measurement is made, and the hydrogen itself must be free from oxygen; this is usually effected by passing the hydrogen over copper turnings maintained at 500°, or over a platinum catalyst at room temperature.

The hydrogen electrode has a number of disadvantages: because oxygen must be excluded from the test solution, all work, especially in alkaline solutions, must be done in closed vessels; other oxidants, including ions of noble metals, must also be excluded, because they may be reduced by the hydrogen gas, and thus undoubtedly cause some variation in the hydrogen ion concentration; similarly, strong reducing agents must also be absent, otherwise the potential of the electrode will be more negative than it should be. A great many substances, such as hydrogen sulphide, hydrocarbons and arsine, poison the electrode. If the hydrogen for the electrode is made from zinc, as is sometimes the case, care must be taken that the metal does not contain arsenic, otherwise some arsine will be carried into the electrode chamber. One further disadvantage is that the composition of solutions containing volatile acids or bases can be changed by the passage of hydrogen, which sweeps them from the solution.

Despite these disadvantages, the hydrogen electrode has been, and still is, used for many applications. The first titration with a hydrogen electrode was reported in 1899, and neutralisation curves for hydrochloric, sulphuric, acetic, boric, carbonic and other acids were given[76]. Hildebrand[69] used a hydrogen electrode for the titration of the salts of weak acids and bases; in addition, he attempted the titration of magnesium with a strong

References p. 165

base, but the results were not accurate. Kolthoff[77] showed that better results were obtained for the magnesium titration at 70°. (The hydrogen electrode can theoretically be used at elevated temperatures, but trouble is sometimes experienced if anything more than a titration curve is desired.)

Several polyprotic acids have been titrated with a hydrogen electrode by Malprade[78], who obtained good results for the titration of phosphoric acid by using calcium hydroxide[79]. The use of hydrogen electrodes for the titration of very weak acids, such as boric acid, has been demonstrated by Mellon and Morris[80].

The hydrogen electrode has been used extensively by Britton and his co-workers for the investigation of various hydroxides and heteropoly acids (for some examples see Reference 81), and it has been recommended for the routine checking of other electrodes and reference solutions[82]. The unique contribution that the hydrogen electrode can make is in the measurement of pH above 12; other electrodes are unusable, or at least unreliable, at such high pH values.

(ii) The Oxygen Electrode

When an unattackable electrode such as platinum is placed in a solution saturated with oxygen or air, the electrode responds to changes in hydrogen ion concentration, presumably because of the occurrence of the following electrode reaction:

$$2H^+ + O_2 + 4e = 2OH^-. \tag{51}$$

The potential of the electrode should then be expressed by

$$E = E^0 - \frac{0.059}{4} \log_{10} \frac{[OH^-]^2}{[H^+]^2 \, P_{O_2}}, \tag{52}$$

but as

$$[OH^-] = \frac{K_w}{[H^+]},$$

this can be shortened to

$$E = \varepsilon + 0.059 \log [H^+] \quad \text{at } 25°, \tag{53}$$

(where ε is a constant independent of pH) provided that steps are taken to maintain the partial pressure of oxygen constant.

Many workers have found this electrode to be irreversible and subject to objectionable drifting, especially in alkaline solutions[83]. Furman[84] found the electrode usable, however, provided that oxygen or air was passed through the solution to ensure saturation. In actual practice, the electrode potential does not always conform to equation (53), but this difficulty has been avoided by empirical calibration in a series of standard buffer solutions[85]. Although the oxygen electrode is subject to drifting and needs empirical calibration, it can be used in solutions containing oxidants, whereas the hydrogen electrode cannot. Not much use has been made of this electrode, but a number of applications were developed in the 1920's; among the most interesting was the titration of nitrophenols, dyes and other compounds by Treadwell and Schwarzenbach[86].

It may be thought that this electrode should be considered with the metal electrodes, such as antimony, which owe their hydrogen ion response to a thin layer of surface oxide, as it has been proved that platinum can become coated with an oxide film (page 87). Richards[87] has shown that a platinum electrode responds to variations in oxygen pressure, indicating that equation (51) probably plays a role in establishing the potential. In some cases (especially in non-aqueous media), the response of a platinum electrode to changes in hydrogen ion concentration can apparently be explained in terms of the presence of an oxide film. In fact, the film must be carefully produced and maintained if the electrode is to function properly[88, 89]. It is evident that the exact mechanism is in some doubt.

(*iii*) *Metal Electrodes*

A number of metal/metal oxide electrodes have been proposed for pH measurements and as indicator electrodes in potentiometric acid-base titrations. In general, response to changes in pH comes about through an equation similar to

$$MO + 2H^+ = M^{2+} + H_2O, \tag{54}$$

provided that the solubility of the metal oxide is small. The relationship between potential and hydrogen ion concentration is then

$$E = \varepsilon + 0.059 \log [H^+] \text{ at } 25°. \tag{55}$$

The antimony electrode is typical; with this electrode, equation (54) becomes

$$Sb_2O_3 + 2H^+ = 2SbO^+ + H_2O, \tag{56}$$

and equation (55)

$$E = 0.050 + 0.054 \text{ pH}, \tag{57}$$

if the potential is measured *versus* the normal calomel electrode at 25°. (The factor 0.054 is empirical.) Franke and Willaman[90], who developed equation (57), indicate that it is useful between pH 0 and 12. Roberts and Fenwick[91] also tested the antimony electrode and found that, with great care, the electrode behaved theoretically, although the attainment of the

References p. 165

theoretical potentials was very slow. The characteristics of the antimony electrode have been thoroughly discussed by Britton[92].

The antimony electrode does not give results of high accuracy, but it is extremely rugged and has a low enough resistance to allow the use of simple potentiometers rather than valve voltmeters. Its main uses are in industrial control and as an indicator electrode for titrations. It must be noted that the antimony electrode does not work well outside the pH range 2 to 9, and is disturbed by strong oxidising or reducing agents and certain anions that form complexes with antimony; anions of hydroxy acids, such as tartrate and citrate, are specially troublesome. The antimony electrode can, however, be used in cyanide solutions.

Britton and Robinson[92] recommend the use of antimony metal bars polished with emery paper, with the addition of some antimony trioxide to the solution. Antimony rods mounted in plastic sleeves are available commercially.

Some of the first acid-base potentiometric titrations with this electrode were developed by Kolthoff and Hartong[94]. Britton and Robinson[93] titrated acetic, phosphoric, boric, malonic, and other acids, as well as tungstates and chromates. Shukoff and Gortikoff[95], using the differential titration technique (two potential jumps), also titrated phosphoric acid, but in an alcoholic solution. They also found that alkaloids could be titrated with an antimony electrode, whereas they interfere with the functioning of a hydrogen electrode. Apparently, the antimony electrode does respond to changes in hydrogen ion concentration in solutions of strong oxidants, but antimony(V) participates in the mechanism of potential establishment, and potential breaks of a different type are found. The antimony electrode has also been used to follow the precipitation of aluminium and magnesium hydroxides from their respective chlorides[96].

The mercury/mercuric oxide electrode has been successfully applied to the titration of ammonia in its salts, of phosphate, and of boric acid[36]. As halides seriously disturb this electrode, it has not been extensively used. The tungsten electrode has been investigated[97] and found to have the serious disadvantage that constant potentials are not attained.

A large number of bimetallic electrode pairs have also been used to find the end points of acid-base titrations; a list of such pairs investigated up to 1942 is given by Furman[62]. Fuoss[51] tested a number of bimetallic electrode systems and found that some of the more useful were the antimony/antimony-lead alloy pair; the bismuth/silver pair and the copper/cupric oxide pair. Kahlenberg and Krueger[98] also tested a number of electrode pairs, and found the silicon/tungsten pair and the tungsten/silver pair to be especially useful.

Harlow *et al.*[88] have used two platinum electrodes for acid-base titrations in non-aqueous solvents, and found that, before each titration, the electrodes had to be polarised in such a way that one was made a cathode and the other an anode.

Very often, the polarisation of an electrode pair increases the rate at which potentials are established and also results in a larger potential jump at the end point. For instance, Wright and Gibson[99] used two similar platinum electrodes between which a current of 20 microamp was passed; the potential of the platinum cathode was then measured against a calomel reference electrode. Sharp breaks were found at, or close to, the equivalence point, when solutions of several acids were titrated with 0.5N sodium hydroxide. Some polarised bimetallic electrode pairs have also been investigated, and the platinum/graphite pair was found to be one of the most useful[100]. Shain and Svoboda[89] have used a polarised platinum electrode pair for the titration of various acids in acetone.

The mechanism by which potentials are established at bimetallic electrode pairs, either polarised or unpolarised, is not entirely clear; it seems reasonable to assume, however, that with platinum electrodes, the cathodised or cathodic electrode behaves, in some sort, as a hydrogen electrode, while the anodised or anodic electrode is a platinum/platinum oxide electrode or an oxygen electrode[89]. Most other electrodes also probably function as metal/metal oxide electrodes. Some theoretical aspects of the participation of oxide films in the measurement of pH have been considered by Hoar[101].

Another type of oxide electrode is the "higher oxide electrode", which has been studied by Parker[102], who found, for instance, that if a platinum wire is placed in contact with solid Mn_2O_3, an electrode that responds to pH changes is obtained; the essential reaction is the equilibrium between Mn(III) and Mn(II).

(*iv*) *The Quinhydrone Electrode*

The quinhydrone electrode is formed by a bright platinum or gold electrode immersed in a saturated solution of quinhydrone[103, 104]. For the best results, the noble metal should be in contact with solid quinhydrone and the solution should be as free from air as possible. Quinhydrone is a slightly soluble molecular compound formed by the addition of one mole of benzoquinone to one mole of benzohydroquinone. As the quinone is more volatile than hydroquinone, commercial quinhydrone should be recrystallised from water at 70°.

The electrode half-reaction is

$$OC_6H_4O + 2H^+ + 2e = HOC_6H_4OH, \qquad (58)$$

and its standard potential is 0.6994 volt at 25°.

References p. 165

The potential of reaction (58) is given by

$$E = 0.6994 - \frac{0.05915}{2} \log_{10} \frac{[H_2Q]}{[Q]\,[H^+]^2} \qquad (59)$$

Because the quinone (Q or OC_6H_4O) and the hydroquinone (H_2Q or HOC_6H_4OH) are obtained by dissolving quinhydrone, $[H_2Q] = [Q]$, and the pH is related to the potential by

$$E = 0.6994 - 0.05915 \text{ pH at } 25°. \qquad (60)$$

Salt errors, which are remarkably small, can be attributed to changes in the activities of H_2Q and Q by different amounts. The magnitude of this error has been recorded for a number of salts[105], but is generally less than 0.04 pH units at a concentration of $1M$. The salt error can be eliminated by saturating the solution with both quinhydrone and hydroquinone, thus fixing the activity ratio of $[H_2Q]:[Q]$; if this is done, the standard potential is altered and a correction must be applied.

Hydroquinone is a weak acid ($K_1 \simeq 10^{-10}$), and quinhydrone electrodes therefore cannot be used in very basic solutions. Above pH 8, the dissociation of hydroquinone becomes appreciable, and the concentration ratio of $H_2Q]:[Q]$ changes from unity. As the pH increases, the slope passes from -0.06 v per pH unit to -0.03 v, and eventually to zero at high pH values[106].

In general, the quinhydrone electrode is somewhat more useful than the hydrogen electrode, as fewer factors interfere with its operation, and it is simpler to construct; such diverse materials as soils[107] and dairy products[108] may be investigated with it. Oxidants and reductants adversely effect the electrode, however, as they react with quinone or hydroquinone. In fact, even the oxygen in the air has a noticeable effect on dilute solutions[109], and an inert gas must be passed through the solution in certain cases. The Q/H_2Q couple has a higher oxidizing potential than the H^+/H_2 couple, so that the former may be used with somewhat stronger oxidants, such as ions of metals more noble than hydrogen.

The quinhydrone electrode has been found to be very useful as an indicator electrode for acid-base titrations[107, 110, 111], because it can be applied to a wide variety of acids and bases of different strengths and concentrations. By using a differential method, Clarke and Wooten[115] were able to titrate acids and bases at concentrations of $4 \times 10^{-4}N$ with an accuracy of 1–2%; it was, however, necessary to perform these titrations in an inert atmosphere in order to avoid oxidation of the hydroquinone and to eliminate errors due to carbon dioxide. The quinhydrone electrode has been used for the determination of the neutralisation values of oils and for the titration of fatty acids[113]; the solvent used was isoamyl alcohol saturated with lithium chloride, and the reference electrode was a second quinhydrone electrode in the solvent.

The constituents of paper have been titrated potentiometrically by Clarke and Wooten[114], who used the Pinkhof-Treadwell technique; the quinhydrone electrode is particularly well suited to this, because the pH values of the equivalence points can generally be calculated with accuracy, and because a large number of buffer solutions of various pH's are available. Extensive studies of the quinhydrone electrode have confirmed its reliability and reproducibility[115, 116].

(*v*) *The Glass Electrode*

A glass electrode consists of a thin-walled glass bulb, which contains an internal reference electrode, usually a silver/silver chloride electrode, in dilute hydrochloric acid. The glass electrode is immersed in the solution to be tested or titrated, and is connected to an external reference electrode, which is usually a saturated calomel electrode, by means of a salt bridge.

As the electrical resistance of the glass membrane is very high and can range from 1 to as much as 500 megohms, some device, such as a valve voltmeter, that will measure potentials through a very high resistance must be used.

Although the first observation[117] that the potential of a glass membrane electrode depends on the acidity of the solution in which it is placed was made in 1906, the first systematic study[118] of the glass electrode was made in 1919; it was not until the 1930's, however, that the glass electrode became widely used. This delay was due to difficulties in the development of suitable electronic measuring instruments and of suitable glasses for membranes. Comprehensive monographs concerning the determination of pH with the glass electrode have been written by Dole[119], and by Bates[19].

The cell used for pH measurements with a typical glass electrode may be represented by

$$\underset{\text{glass}}{\text{Ag/AgCl(s), HCl}} \text{ / test solution / saturated calomel electrode.} \tag{61}$$

At 25°, the potential of this cell obeys the equation

$$E = \varepsilon + 0.05915 \text{ pH} \tag{62}$$

over the pH range of 0 to 11. The value of ε depends on the reference electrode and on the values of the various junction potentials. The junction potential at the glass membrane is generally quite large, and is often termed the asymmetry potential. The asymmetry potential is more exactly defined as the potential measured when identical reference electrodes are placed on either side of a glass membrane, both sides of which are bathed in the same

References p. 165

buffer solution. This asymmetry potential apparently results from differences in the condition of the two surfaces of the glass resulting from strains introduced during fabrication, unequal chemical attack, and other causes. Because the asymmetry potential of any particular glass electrode is continually changing, frequent calibration by means of standard buffer solutions is necessary.

The exact mechanism by which a glass electrode functions is not completely understood. One reasonable explanation is that the glass electrode acts as a semi-permeable membrane through which only hydrogen ions can pass; hydrogen ions do appear to be transferred through the glass membrane by the passage of current in accordance with Faraday's law; it is also known that the presence of water in the glass is essential for the functioning of the electrode. For this reason, glass electrodes should be stored in water when not in use, and a dry electrode must be soaked in water for two or three hours before it will give constant potentials. Glass electrodes should not be placed in dehydrating media unless absolutely necessary. Thus, if non-aqueous titrations are to be performed with a glass electrode, it is best to soak the electrode in water between experiments. When necessary, a glass electrode is best cleaned in dilute hydrochloric acid, and water, and/or wiped gently with soft tissue.

Sometimes, if there is an exceptional lag in the response of the electrode, or if it cannot be standardised in the range of the pH meter's asymmetry control, a glass electrode may be rejuvenated by soaking it in a 20% solution of ammonium bifluoride for about three minutes.

Various types of soft glasses are used for the construction of glass electrodes. MacInnes and Dole[120, 121] studied many different types and found that a soda/lime glass was particularly useful. A glass based on their findings is now often used (Corning 015: 72.2% SiO_2, 21.4% Na_2O and 6.4% CaO). Perley[122] has also extensively studied the effect of the composition of the glass on pH response and resistance.

Corning 015 glass shows a departure from equation (62) in solutions whose pH is above 11, especially if large amounts of sodium or other alkali metal ions are present. In general, the "alkaline error" is negative, in the sense that the observed pH is too small by as much as one pH unit, under especially bad conditions. Perley[122] showed that this error could be reduced by the use of a lithium glass, and special electrodes of this type for measurement of strongly basic solutions are available from most makers.

The glass electrode is the most useful of all pH electrodes; it can be used in the presence of strong oxidants or reductants, in very viscous solutions, and in the presence of proteins and alkaloids, which often seriously interfere with other types of electrodes. Glass electrodes, together with the proper, readily available, electronic equipment, can easily be adapted to automatic titrations and the recording of pH. Micro glass electrodes are also available for micro titrations[123] involving volumes as small as one drop.

Because the response of glass electrodes is so nearly theoretical, the end point of almost any acid-base titration can be found, provided that the slope of the titration curve at the end point is steep enough to establish the end point accurately (Section 3 *b*). Of the more interesting acid-base titrations with a glass electrode, the determination of metal ions by the titration of hydrogen ions released by a complexing reaction, such as that with nitrilotriacetic acid[124] for example, may be mentioned:

$$M^{2+} + HX^{2-} = MX^{-} + H^{+},$$

where X^{3-} is the nitrilotriacetate ion; another complexing agent that can be used is EDTA. When iron, aluminum, or chromium are titrated, two potential jumps are observed, owing to the fact that these metal ions form strong hydroxy complexes; titrations of this type have been summarised by Schwarzenbach[125].

Another method for the determination of metal ions by means of acid-base titrations with a glass electrode is exemplified by the work of Day *et al.*[126] who passed metal ion solutions through an ion-exchange resin in the hydrogen form, thus converting the metal ions quantitatively into hydrogen ions, which were then titrated with a standard base solution.

(*vi*) *Non-aqueous Titrations*

The potentiometric detection of the end points of acid-base titrations has been extensively applied to non-aqueous solvents, and almost all the indicator electrodes previously discussed have been used; most of the work has been done with metal electrodes (especially antimony), or with the glass electrode. Care must be taken when the glass electrode is used, because water plays an important role in the mechanism of its pH response; intermittent soaking in water usually permits the successful application of the glass electrode. Reference electrodes must be so chosen that the total resistance of the cell does not become so great that measurements are impossible; the asbestos-fibre type of calomel electrode is usually avoided in non-aqueous work, and is replaced by the sleeve type, or by some other electrode of low resistance.

Most non-aqueous potentiometric titrations are carried out empirically and, even when the glass electrode is used with a standard pH meter, millivolts are recorded instead of pH values, because the meaning of pH in non-aqueous media is not well understood. Kolthoff and Bruckenstein[127, 128] have, however, made considerable progress toward putting non-aqueous titrations on a sound theoretical basis.

References p. 165

Some of the techniques and applications of non-aqueous potentiometric titrations have been discussed by Fritz[129] and are also considered in Vol. I B, p. 796; the choice of solvents for various titrations has been dealt with extensively by Fritz[130] and van der Heijde[131].

(b) Precipitation Titrations

A large amount of work has been done on the potentiometric detection of the end points of precipitation titrations; this is because not many indicators are available, and those that are cannot often be used successfully with mixtures. The main classes of reactions that will be considered are those whose end points can be detected with a silver, a mercury, or a platinum indicator electrode. The precipitation of metal hydroxides can be followed by many of the acid-base indicator electrodes discussed in the previous section, but the titration of metal ions with a standard base is usually not successful, because of their great tendency to form mixed salts of various types.

(*i*) *Silver Electrodes*

A silver electrode usually consists of a piece of pure silver wire, or of a platinum wire or gauze plated with silver. If a typical calomel reference electrode is used in conjunction with the silver indicator electrode, contact between the former and the solution must generally be effected by means of a potassium nitrate salt bridge; this avoids the possibility of the solution becoming contaminated with chloride and mercurous ions. In the interest of simplicity, some other reference electrode may be chosen, such as a glass electrode, provided that a valve voltmeter is used as a measuring device. The use of calomel electrodes of the fibre type has been recommended[132], because the diffusion of chloride through the tip is relatively slow (0.005 millimoles of chloride in four hours.)

When a silver electrode is used with extremely dilute solutions, it is found that better results are obtained if air is displaced by the passage of an inert gas. Apparently, oxygen can effect the potential of a silver electrode, if it is not well stabilised by a significant amount of silver ion or halide ion. In iodide titrations, small amounts of iodide may be oxidised to iodine by the oxygen of the air; this reaction depends on the acidity of the medium and can generally be avoided by working with neutral solutions.

The best known titration with a silver indicator electrode is the determination of halides, originated by Behrend[133], who determined chloride, bromide, and iodide with an accuracy of about 0.5%. Treadwell[134] has investigated the influence of extraneous salts on the halide titration: because of the formation of complexes and changes in activities, the presence of large concentrations of salts often adversely affects the accuracy of halide titrations; errors may be avoided by the use of the Pinkhof-Treadwell method, provided that the reference solution has the same composition as the test solution at the equivalence point, but this is difficult to achieve in practice.

Adsorption of various types is also a serious difficulty and it has been found that the effect of adsorption is more pronounced the more insoluble the precipitate[135]; this difficulty is particularly troublesome in the analysis of mixtures.

Lange and Schwartz[136] have proved that the potentiometric titration of chloride and bromide can be carried out with extreme accuracy: by using weight burettes and low temperatures, and by being careful to provide very thorough stirring, they attained an accuracy of 0.007%; contamination of the solution was prevented by connecting the reference electrode to the solution by means of an ammonium nitrate salt bridge. Without special precautions, except the use of a potassium sulphate salt bridge, Kolthoff and van Berk[137] were able to carry out the same titrations with an accuracy of 0.02%.

Iodide ion has been determined in extremely dilute solutions[138], but steady potentials were obtained only after prolonged periods. Lange and Berger[139] determined iodide at a concentration of only 0.005 mg per litre, and they noticed that the potential depended on the rate of stirring; the reason for this is not known, but is has been suggested that the precipitate, by rubbing on the electrode, creates an electrostatic charge.

The accurate determination of mixtures of halides has been the goal of many workers. The titration of iodide in the presence of chloride and bromide only gives acceptable results if the solution is extremely well stirred; otherwise, the formation of a mixed halide precipitate results, with large positive errors in the iodide figure. The determination of the sum of all three halides presents no problem.

Clark[140] attempted to prevent the formation of mixed halides by the addition of extra salts, and found barium nitrate to be especially useful. He also devised a method, which is described briefly below, to prevent the interference of iodide and bromide with the chloride titration; this involves the destruction of the bromide and iodide, and was applied to the determination of chloride in potassium bromide.

Determination of Chloride in the Presence of Large Amounts of Bromide or Iodide

Dissolve a 2.5 g sample in 350 ml of distilled water and add 25 ml of concentrated sulphuric acid. Boil the solution, and then add 0.5N potassium perman-

References p. 165

ganate solution until a permanent violet colour is obtained. Boil the solution again until the bromine is removed. Add 10 g of ammonium nitrate, and titrate the chloride potentiometrically with 0.01 *N* silver nitrate solution.

A similar procedure has been devised by Tomicek[141].

Martin[132] has studied the problem of the determination of mixtures of halides by means of a single potentiometric titration with a standard silver solution; as stated previously, the total halide can be determined accurately,

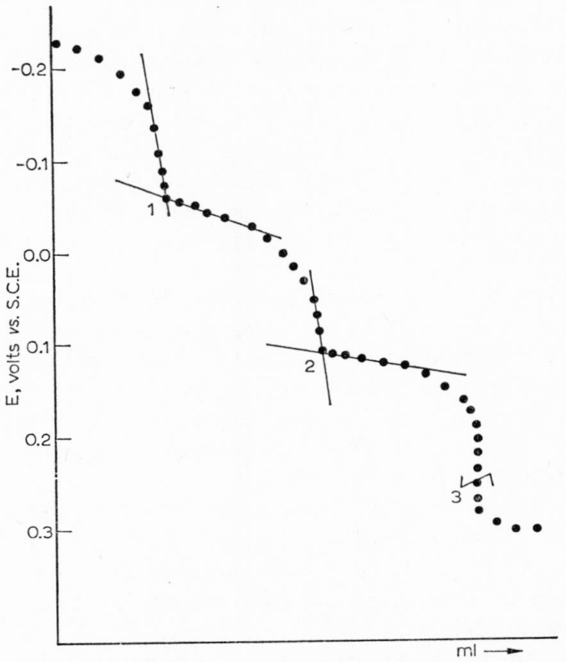

Fig. III, 22. Potentiometric titration of halide mixtures.

but individual halide determinations are generally in error because of mixed halide precipitation. Martin's method involved the development of an empirical correction to be applied when the mole ratio of halide mixtures was greater than 1 to 20; for smaller ratios, more accurate results were obtained without applying the correction. The correction greatly reduces the errors found in the iodide and bromide end points because of mixed halide formation. Another important feature of this method is that the iodide and bromide end points are obtained from the intersection of two straight lines (see Fig. III, 22) rather than by one of the usual methods such as calculation of the maximum value of dE/dV or titration to some specific pre-calculated potential; in this particular determination, the straight line method appears to give good results, especially when the correction is applied, but it is not recommended for general use. Martin's method is as follows:

Potentiometric Titration of Halide Mixtures[132]

Apparatus:

Any pH meter, to which are connected a silver wire electrode and a fibre-type of saturated calomel electrode, which must have a very low leakage rate.

Procedure:

Pipette the sample solution into a 400 ml beaker and dilute it to 150 ml with water (see Note below). Add $1.5M$ nitric acid to bring the solution to about pH 1 as shown by a suitable indicator paper; avoid the addition of too much nitric acid as this may oxidise the iodide ion to iodine. Titrate the solution potentiometrically and record as many points as convenient, especially near the inflection points. Determine the first two end points by drawing two straight lines through the two branches of the curve on either side of each intersection point; take the final end point as the maximum value of dE/dV. Calculate the amounts of iodide, bromide and chloride from the following equations:

$$(\text{mmole}\,I^-)_{\text{corr.}} = 0.9908\,(\text{mmole}\,I^-)_f + 0.0016\,(\text{mmole}\,Br^-)_f + 0.0023\,(\text{mmole}\,Cl^-)_f + 0.0019$$

$$(\text{mmole}\,Br^-)_{\text{corr.}} = 0.0051\,(\text{mmole}\,I^-)_f + 0.9831\,(\text{mmole}\,Br^-)_f - 0.0036\,(\text{mmole}\,Cl^-)_f + 0.0091$$

$$(\text{mmole}\,Cl^-)_{\text{corr.}} = (\text{mmole total halide})_f - (\text{mmole}\,I^-)_{\text{corr.}} - (\text{mmole}\,Br^-)_{\text{corr.}},$$

where the subscript "f" signifies the quantities found by experiment, and the subscript "corr." the corrected values.

Note:

For the constants in the above equations to be valid, the experimental conditions—*e.g.*, the total volume of titrated solution—must be reproduced accurately; if this is done the results are accurate to about 1% or less.

Potentiometric end-point detection has also been applied to coulometric halide titrations with electro-generated silver ion[142, 143].

The end points of titrations involving a wide variety of insoluble silver salts have also been found by using a silver indicator electrode: in the determination of thiocyanate[144], barium nitrate is added to prevent adsorption errors; copper and cadmium have been determined by precipitating such salts as $Cd\,py_2(CNS)_2$, where py stands for pyridine[145]; the direct titration of ferrocyanide with a standard silver solution is also possible, but it should be noted that two potential jumps occur[146], the first being due to the formation of $KAg_3Fe(CN)_6$, and the second to $Ag_4Fe(CN)_6$.

Thiosulphate has been titrated by Müller[147], who found that the titration curve indicated the formation of the complex anion $AgS_2O_3^-$ followed by the precipitation of $Ag_2S_2O_3$; the potential break for the precipitation is the larger and more useful. If the titration is not carried out at 0°, the silver thiosulphate decomposes to silver sulphide, and the difficulties involved make the titration impracticable for analytical work. Because of its solubility, silver chromate cannot be used to determine chromate[145], but phosphate and arsenate can be determined by precipitation of their silver salts[37]. Careful control of the pH is necessary for accurate results, and some workers[149] have found it necessary to maintain the pH of this solution at 9 by the continuous addition of alkali during the titration.

In the titration of sulphide, Dutoit and von Weisse[150] encountered errors as large as 10% with a polarised electrode system, though Kolthoff and Verzijil[5] obtained good results with a silver indicator electrode, provided that they used fairly concentrated solutions.

The reaction between sodium nitroprusside and silver nitrate has been investigated potentiometrically by Scagliarini and Pratesi[151]. Formaldehyde can be determined by the addition of excess of silver ion to a formaldehyde solution containing carbonate:

$$HCOH + 2\,Ag^+ + H_2O = HCOOH + 2H^+ + 2Ag;$$

the excess of silver ion is back-titrated with potassium chloride solution[152].

The titration of mercaptans in the presence of sulphides has been studied by Davies and Armstrong[153], and the reaction has been extensively used in the analysis of gasoline and other petroleum products. Karchmer[154] has done much to improve the accuracy of this titration by pointing out that inorganic sulphides may be formed in alkaline alcoholic solutions; decreasing the alkalinity and removing air reduces the extent of polysulphide formation. Coulometric titrations have been extensively applied in this field.

The salts of several organic acids such as lauric, myristic, palmitic and stearic have been titrated potentiometrically[155]. Perchlorates and similar salts have been determined potentiometrically after reduction by refluxing with aluminium metal and tervalent titanium[156].

A complex mixture containing sulphide, cyanide, thiocyanate and chloride ions has been determined by titration with silver[157]. Four potential breaks are obtained: the first corresponds to the precipitation of silver sulphide, and the second to the formation of $Ag(CN)_2^-$; the third break occurs after both silver cyanide and silver thiocyanate have been precipitated, and the final potential jump indicates that all the chloride in solution has been converted to silver chloride.

Interest in the sodium salt of tetraphenylborate (TPB) has led to the development of a number of potentiometric methods based on the fact that the silver salt of TPB is insoluble. Crane[158] has determined a number of amines by precipitating the corresponding tetraphenylborate with an excess of TPB; the precipitate is filtered and washed with a 1:1 acetone-water mixture, and the excess of TPB is titrated potentiometrically with standard silver solution. In this method, the silver solution must be standardised against TPB; otherwise, errors result. An extensive study by Kirsten *et al.*[159] has shown that the direct titration of many organic and inorganic bases is possible with a silver/calomel or a silver/glass electrode pair. Potassium, rubidium and caesium sulphates were titrated with good results as was the drug, "Oxadition".

It was also found that a positively polarised silver electrode gave a sharper end point than an unpolarised one. Various other electrode pairs were investigated, including a silver/silver and platinum/platinum polarised pair. All solutions were buffered at about pH 4.5. Halides were found to interfere with the detection of the end point, but this difficulty was avoided in many cases by replacing the halide with acetate by means of an anion-exchange column.

The silver electrode may be used in the determination of silver, as well as for the titration of many anions with standard silver nitrate solutions. Hilter and Gittel[160] have developed a systematic scheme of potentiometric precipitation titrations for the analysis of mixtures containing Ag, Bi, Pb, Cu, and Cd. Silver has been titrated with iodide in the presence of a number of other metals such as Cu, Pb, Bi, Cd, Zn and Fe[161], which are masked by complex formation with EDTA. The titration of silver, copper and mercury(II) with dithio-oxamide has also been followed potentiometrically[162].

One of the most important applications of the silver electrode is the determination of silver in photographic materials, and Schulze[163] has described a rapid titration of silver with a standard sodium sulphide solution in which the end point was detected by a differential method: two silver/silver sulphide electrodes were used, one placed in the direct path of the titrant as it issued from the burette, the other a little outside the titrant stream; by controlled stirring, a concentration cell was thus set up in the solution, and a large potential difference was obtained in the vicinity of the end point. The titration of silver with thioacetamide has also been applied to the analysis of photographic materials[164]: when thioacetamide is added to a solution of silver, silver sulphide is precipitated almost instantaneously, even at room temperature, and because it is so insoluble, a large potential break is indicated by a silver electrode; the method is as follows:

Determination of Silver

Apparatus:

A pH meter or automatic titrator equipped with a silver/silver sulphide electrode (Note 1) and a calomel electrode.

Reagents:

Buffer, pH 5. Mix 0.1M potassium acid phthalate and 0.05M trisodium potassium phosphate solutions in the ratio of 50 to 24 parts by volume, and add 0.5 g of powdered thymol per litre (Note 2).

Thioacetamide solution. 0.2 N stock solution. Dissolve 7.6 g of thioacetamide in 1 litre of the buffer solution of pH 5

Thioacetamide solution. 0.01 N standard solution. Dilute 50 ml of the 0.2N stock solution to 1 litre with the buffer solution of pH 5. To standardise the reagent, add 10 ml of 0.02N potassium bromide to 10 ml of 0.01N standard silver nitrate solution, add sodium thiosulphate, gelatine and EDTA solutions as directed below, and titrate the liquid with the 0.01N thioacetamide.

Procedure:

To an amount of the sample (liquid or film) known to contain about 5 mg of silver, add 10 ml of a 24% solution of sodium thiosulphate, followed by 5 ml of a 0.4% solution of gelatine and 25 ml of 0.4% solution of EDTA (Note 3) in 2M sodium hydroxide. Titrate the mixture potentiometrically with 0.01N thioacetamide to the inflection point of the titration curve, using the calomel reference electrode and the silver/silver sulphide indicator electrode.

Notes:

(1) A silver electrode may be given a sulphide coating by soaking it in a 20% solution of sodium sulphide for 3 minutes.

(2) The thymol is added as a bactericide; at pH 5 and with thymol present, the reagent is virtually stable for at least 40 days.

(3) The EDTA is added to prevent the interference of other metal ions such as Cu, Cd, Zn and Ni.

This method can also be applied to mercury(II), interference from other metal ions being removed by the EDTA; common anions, except large amounts of cyanide, do not interfere.

(*ii*) *Mercury Electrodes*

A mercury electrode is a useful indicator electrode for precipitation titrations involving some species of mercury. One form of this electrode is a mercury pool at the bottom of the titration vessel or in a small cup at the end of a J-shaped glass tube; contact with the pool is generally made by a small platinum wire sealed into the wall of the glass container. A more convenient type of electrode consists of a noble metal, such as platinum or gold, coated with mercury; the mercury coating may be applied physically or by electroplating.

If a mercury electrode is to be used in a titration involving mercuric ion, the fact that the following reaction occurs must sometimes be considered:

$$Hg + Hg^{2+} = Hg_2^{2+}; \quad K = 166.$$

In most practical work, this reaction is not of importance, because excess of mercuric ion is not present in appreciable amount. In any event, the potential of a mercury electrode is influenced by both mercuric and mercurous concentrations.

A number of titrants have been recommended for use with mercury electrodes: mercurous nitrate containing a little nitric acid is a good source of mercurous ions, but mercurous perchlorate is also often used. Mercurous perchlorate may be prepared by heating pure mercuric oxide in perchloric acid over mercury. Mercuric nitrate, perchlorate and chloride have been recommended as titrants containing mercuric ion.

Mercurous nitrate has been used to titrate chloride and bromide[5, 165], but mercurous perchlorate was of no use for these determinations[166]. In addition, although the end point of the iodide titration is not well defined because of complex formation, the titration of mercurous perchlorate with iodide is somewhat better. Iodide is best titrated with mercuric chloride, mercuric iodide being formed[5]; some complex formation between mercuric ion and iodide occurs, but the potential jump corresponding to the formation of HgI_4^{2-} is too small for analytical purposes. Chloride and even bromide do not interfere, provided that their concentrations do not exceed that of the iodide. Thiocyanate has been determined with mercuric nitrate and perchlorate[5].

The titration of oxalate with a weakly acidic solution of mercurous nitrate has been made the basis of the determination of several metals such as calcium, strontium, cadmium and lead[167]. An excess of a standard oxalate solution is added to the sample and back-titrated. Sulphide can be titrated with mercuric chloride in excess of sodium hydroxide.

The determination of alkaloids has been studied extensively by Maricq[168], who precipitated the alkaloid with an excess of potassium iodomercurate and then titrated the unused iodomercurate potentiometrically with mercuric chloride and a mercury electrode. The precipitation reaction is as follows:

$$H_2HgI_4 + A^{2-} = HIA(HgI_2) + HI,$$

where A is the alkaloid being determined. Maricq's method is given in detail below.

References p. 165

Determination of Alkaloids

Reagent:

Potassium iodomercurate. Saturate a 0.025*M* potassium iodide solution with mercuric iodide and filter off the excess; standardise the reagent by titration with 0.005*M* standard mercuric chloride solution.

Procedure:

Add 10 ml of the reagent to a sample containing 10–40 mg of alkaloid dissolved in 10 ml of 0.1*N* sulphuric acid; a precipitate should form. Stir the solution for three minutes and filter it; the filtrate should be clear. For each 10 ml of filtrate, add 35 ml of water and 10 ml of 1*N* nitric acid, and titrate the solution potentiometrically with 0.005*M* mercuric chloride solution, using a mercury electrode to detect the end point.

In a few cases, amalgam electrodes have been found useful; for instance, the titration of sulphate with lead or barium solutions has been followed with lead amalgam electrodes[169].

Mercury indicator electrodes have been used to detect the end points of titrations involving electro-generated mercurous or mercuric ions; the halides[170, 171] have been investigated as well as sulphide[172].

(*iii*) *Iodine Electrodes*

An iodine electrode consists of a piece of bright platinum wire or gauze in contact with iodine. The potential of such an electrode is governed by the relationship[173]

$$E = \varepsilon + 0.017 - 0.59 \log [I^-]. \qquad (63)$$

Thus, the electrode responds to variations in the iodide-ion concentration, and can be used for titrations involving the precipitation of insoluble iodides. Actually, the potential jump at the end point depends on the hydrogen ion concentration, because the reaction

$$3I_2 + 3H_2O = 6H^+ + IO_3^- + 5I^- \qquad (64)$$

can occur if iodide ion is removed from the solution; the largest potential jumps are found in acid solution.

Iodide ion, down to a concentration of $10^{-3}M$, can be determined with an accuracy of 1% by titration with a standard silver nitrate solution. The presence of chloride or bromide causes the potential break at the end point to be smaller. Mercuric perchlorate may also be used as a titrant[173]. In this titration, bromide in amounts equal to the iodide causes no interference and the presence of chloride causes no difficulty. Actually, chloride and bromide may be titrated as well, but a rather small change in potential is found.

(*iv*) *Platinum Electrodes*

An ordinary platinum indicator electrode may be used to find the end point of certain precipitation titrations in which one of the species involved must be capable of establishing a more or less reversible potential at a platinum electrode: for instance, Atanasiu[174] has used a platinum indicator electrode for the precipitation titration at 70° of barium, lead and mercury as chromates in a medium consisting of 30% alcohol and 70% water; the presence of acids decreased or eliminated the potential break.

The largest class of precipitation titrations with platinum indicator electrodes involves the precipitation of ferrocyanides, and the potential at the platinum electrode is established by the ferrocyanide/ferricyanide couple. To ensure steady potentials, a small amount (3–4 drops of a 1% solution) of potassium ferricyanide is added to the solution to be titrated. Standard solutions of potassium ferrocyanide are reasonably stable, provided that they are stored in dark glass bottles; this stability is improved if they contain 0.2% of sodium carbonate.

The titration of zinc[175] with ferrocyanide has been extensively investigated by many workers. Lingane and Hartley[176], who have shown the importance of pH control in this titration, found that the optimum conditions for the precipitation of $K_2Zn_3[Fe(CN)_6]_2$ lay between pH 2 and 3; not only did a stoichiometric reaction occur, but steady potentials were obtained rapidly; interference due to many metal ions can be eliminated by the addition of EDTA[177]. Cadmium has also been titrated with ferrocyanide, but the results are generally not accurate[175, 178]. The titration of lead[179] is possible, and can be applied to the indirect determination of sulphate. Ferrocyanide has also been used for the determination of copper[180]; in this reaction, ammonia and an oxidising agent such as potassium iodate must be added. Thallium can be titrated with ferrocyanide in the presence of calcium[181] to give $CaTl_2[Fe(CN)_6]_2$.

It is often true that the presence of other salts greatly effects the results of ferrocyanide titrations. This is partly due to pH changes, which vary the predominant species of hydroferrocyanic acid in solution[176]; the presence of various cations can, however, influence the solubility and the form of the precipitate, as is evident from the presence of calcium in the thallium precipitate mentioned above, and from the presence of potassium in zinc ferrocyanide precipitates. Because of these difficulties, the results of ferrocyanide precipitation titrations are not always accurate, especially if the sample contains a large amount of salt.

The use of lead nitrate as a precipitating agent has been recommended for the potentiometric titrations of sulphate, chromate, carbonate, sulphite, tungstate and molybdate[182, 183].

References p. 165

Polarised platinum electrodes have been used for the potentiometric end-point detection of the titration of chloride with silver nitrate[184].

(v) *Third Class Electrodes*

An electrode of the type

$$\text{Metal/M'A(solid),M''A(solid)},$$

where M′ and M″ stand for metals whose salts with A are slightly soluble, may be used for the end-point detection of certain precipitation titrations. The electrodes that have been investigated have mostly been those that might be used for the titration of calcium, *e.g.*

$$Ag/Ag_2C_2O_4,\ CaC_2O_4/\text{Reference electrode.}$$

The theory of such electrodes has been developed by LeBlanc and Harnapp[185], who conclude that the ideal condition for a third class electrode is attained when

$$\frac{2S_1 + S_2}{2(S_1 + S_2)} = 0.5, \qquad (65)$$

where S_1 and S_2 are the solubilities of the two insoluble salts involved. Table III, 4 lists electrodes of this type that have been used.

TABLE III, 4

THIRD CLASS ELECTRODES

Electrode	*Substance Determined*	*Reference*
$Ag/Ag_2C_2O_4$, CaC_2O_4, Ca^{2+}	Calcium	186
Pb/PbC_2O_4, CaC_2O_4, Ca^{2+}	,,	186
$Ag/Ag_3(PO_4)_2$, $Ca_3(PO_4)_2$, Ca^{2+}	,,	186
$Pb/Pb(IO_3)_2$, $Ca(IO_3)_2$, Ca^{2+}	,,	186
Pt/CuC_2O_4, CaC_2O_4, Ca^{2+}	,,	187
$Hg(Pb)/PbC_2O_4$, CaC_2O_4, Ca^{2+}	Calcium in presence of Cl^-	187, 188
$Pt(Hg)/HgS$, PbS, Pb^{2+}	Lead	189

(vi) *Membrane Electrodes*

A membrane made of the substance being precipitated can often serve as an indicating electrode. The membrane is generally sealed across the end of a glass tube, inside which a reference electrode is placed; the assembly is then immersed in the solution to be titrated, together with a second reference electrode, and the potential between the two reference electrodes is measured.

The arrangement is similar to that used with a glass (membrane) electrode.

Silver halide membrane electrodes have been used for the detection of the end points of the titrations of chloride, bromide, iodide and sulphide with silver nitrate[190, 191]. Membranes that respond to calcium ion have been reported[192], and barium membranes have been prepared[193]; the latter have been found to be specially useful in the titration of sulphate with a solution of a barium salt. Membranes made of ion-exchange resins have also found use in potentiometric titrations[194].

The use of membranes that exhibit potentials due to different concentrations of specific ions on either side will doubtless increase in the future as mechanical difficulties are overcome.

(c) Complexometric Titrations

In a number of determinations, complexometric titration can be used instead of precipitation titration; other things being equal, complexometric titrations have the advantage that there are no difficulties due to co-precipitation phenomena or to the fouling of the indicator electrode by particles of precipitate. Complexometric titrations involving cyanide ion have been widely used for many years, and titrations with a chelating agent such as ethylenediaminetetra-acetic acid (EDTA) have now come into prominence because of their wide applicability, and because they make possible the determination of mixtures of various metal ions.

(i) Silver Electrodes

The silver electrode is especially useful as an indicator electrode for the titration of cyanide with silver ion. In this titration, the main potential change is found at the point where $Ag(CN)_2^-$ is completely formed. The strength of the silver cyanide complex is so great (the dissociation constant is 10^{-21}) that cyanide may be determined in the presence of ferrocyanide[195] and the halides[196]. In fact, when a mixture of the halides and cyanide is titrated, three potential breaks occur: the first corresponds to the formation of the silver cyanide complex, the second to the precipitation of silver iodide, and the third to the precipitation of silver cyanide, silver bromide, and silver chloride. The cyanide titration can also be applied to photographic materials such as film; it is possible to dissolve the silver iodide and/or bromide in an excess of cold standard cyanide solution and back-titrate the excess with a standard silver solution. Gelatine from the emulsion may have a disturbing effect, but it can be destroyed by ignition with sodium carbonate.

Halogens may be determined by reduction with zinc, precipitation with silver ion, and titration of the precipitate with cyanide[197]; potential breaks are obtained for the dissolution of each silver halide.

References p. 165

The theory of the argentometric titration of cyanide has been discussed by Ricci[198], who investigated the curve of p[Ag^+] to determine the position of the inflection point in relation to the equivalence point; he also considered the effect of ammonium hydroxide and the precipitation of silver cyanide, tabulated the titration errors at different concentrations and gave equations for the calculation of equilibrium constants from the characteristics of the titration curve.

The determination of various metals that form strong cyanide complexes can be effected by a back-titration method. Nickel forms a cyanide complex indicated by the following reaction:

$$Ni^{2+} + 4CN^- = Ni(CN)_4^{2-}$$

If an excess of standard cyanide solution is added to a solution containing nickel, the concentration of the nickel may be found by titration of the excess of cyanide with a standard silver nitrate solution[199]; the titration is not exact, but it gives better results than the direct titration of nickel with cyanide; the accuracy of the back-titration is improved by the addition of tartrate or citrate[200]. Other workers have also studied and improved this titration[201, 202].

Cobalt may be similarly determined: if an excess of standard cyanide solution is added to the solution containing cobalt and back-titrated with silver nitrate[203], the potential jump occurs at a point corresponding to the formation of $Co(CN)_5^{3-}$. The back-titration with silver is accurate, although the direct titration of cobalt with cyanide is not. Zinc can also be determined by a back-titration method[204], but solutions more dilute than $5 \times 10^{-3}M$ must be avoided or significant errors will be caused by hydrolysis.

(ii) Platinum Electrodes

A bright platinum electrode coupled with a calomel reference electrode can often be used to detect the end point of complexometric titrations. For a sharp end point to be obtained, some reversible oxidation-reduction couple must be present, and one member of this couple must form strong complexes with the complexing agent being used. It is not necessary, however, for the substance being analysed to be one member of the reversible couple.

Aluminium has been titrated with a standard solution of sodium fluoride, the titration medium being a mixture of equal volumes of alcohol and water saturated with sodium chloride[205, 206]. A small amount of ferrous and ferric iron is added in order to establish a steady and useful potential at the platinum indicator electrode; when the aluminium has been completely complexed with fluoride, subsequent additions of titrant will complex the ferric iron and cause a jump in the potential at the indicator electrode. Fluoride may also be used to titrate uranium(IV) according to the reaction

$$U^{4+} + 5F^- = UF_5^-.$$

In this case, the platinum electrode responds more or less reversibly to the uranium concentration, and the addition of an extraneous oxidation-reduction couple is unnecessary[207].

The platinum/calomel electrode system has been applied to complexometric titrations with EDTA. Přibil *et al.*[208] determined Al, Cu, Cd, Ni, Zn, Pb and Bi by adding an excess of standard EDTA solution and back-titrating with standard ferric solution; back-titration was necessary to ensure the existence of a reversible couple to establish a potential at the indicator electrode[209]. The complexometric determination of mixtures of iron, aluminium, and chromium is also possible with a platinum indicator electrode; the analysis of the individual components is based on the fact that chromium and especially aluminium form EDTA complexes very slowly under certain conditions. The determination of copper in an ammonium acetate solution by titration with EDTA and a platinum indicator electrode has also been studied[210]. Zinc, cadmium and nickel in electroplating baths have also been directly titrated with EDTA[211].

Siggia *et al.*[212] have made an extensive study of the potentiometric detection of the end points of titrations of metal ions with chelating agents, including EDTA: they found that the platinum/calomel combination worked well for the direct titration of ferric and cupric ions; a silver indicator electrode could be effectively used in the titration of mercury; most other metal ions could be accurately titrated only with a platinum electrode covered with a coating of mercury; a mercury-pool electrode would undoubtedly be just as effective, but is generally not as convenient.

The platinum/tungsten electrode pair has also been used for end-point detection in EDTA titrations[213]: the titration of lead is characterised by a sharp potential change; other metals such as nickel and copper can be determined by the addition of excess of EDTA, followed by titration of the excess with a lead solution. The titration of the alkaline earth metals with the platinum/tungsten electrode pair, as well as with the tungsten/molybdenum pair, has also been studied[214].

Constant-current potentiometry with one polarised platinum electrode, the potential of which is measured against a saturated calomel electrode, has also been applied to complexometric titrations. The theory of constant-current potentiometric EDTA titrations has been discussed by Kolthoff[215]. Copper has been titrated with EDTA[216] by using a rotating platinum electrode in a 0.5M sodium acetate solution of pH 5, or in ammonium hydroxide[217];

References p. 165

polarising currents of two to five microamp are suitable for these determinations. The form of the curve obtained is typical of a "reversible" substance being titrated with an "irreversible" reagent: before the end point, the indicator electrode (cathode) shows a relatively positive potential, because hydrated or weakly complexed cupric ion is being reduced at the electrode surface, and the exact value of the potential depends on the titration medium used; at the end point, the indicator electrode potential moves to a more negative value, corresponding to the reduction of the copper/EDTA complex or of hydrogen ions, according to the pH. The end point of the titration of copper with α-benzoinoxime has also been detected by constant-current potentiometry[217].

(*iii*) *Mercury Electrodes*

The mercury electrode has also been used for a variety of complexometric titrations; mercury-pool electrodes and platinum or gold electrodes coated with mercury have both been used to advantage. Siggia, Eichlin, and Rheinhart[212] have shown that platinum coated with mercury can be used as an indicating electrode for the titration with EDTA of Cu, Zn, Pb, Mn, Ca, Mg and Ni; complexometric reagents such as *N,N'*-di(β-hydroxyethyl)-glycine and nitrilotriacetic acid were also investigated. Reilley has indicated the general usefulness of the mercury electrode by pointing out that it may be used to detect the end points of the EDTA titrations of some thirty different metal ions[218]. The mercury electrode has also proved useful in ultramicro titrations involving chelating agents[219].

As used by Reilley and his co-workers[220, 221, 222], the mercury indicator electrode may be represented by

$$Hg/HgY^{2-n}, MY^{2-n}, M^{2+}, \tag{66}$$

where Y^{2-n} stands for the completely dissociated chelon (*e.g.* EDTA) and M^{2+} is the metal being titrated. In order to set up a half-cell of this type, a small quantity of mercury(II)-chelonate, HgY^{2-n}, is added to the solution being titrated; the potential of the electrode is given by the expression

$$E = E^0{}_{Hg^{2+}/Hg} + 0.0296 \log_{10} \frac{[M^{2+}]\,[HgY^{2-n}]}{[MY^{2-n}]\,K_{HgY}} K_{MY} \tag{67}$$

at 25°. Equation (67) is arrived at by combining the Nernst equation for a mercury/mercuric ion electrode and the stability constants for the complexes involved. For example,

$$K_{HgY} = \frac{[HgY^{2-n}]}{[Hg^{2+}]\,[Y^{-n}]} \tag{68}$$

Equation (67) shows that there is a linear relationship between log $[M^{2+}]$ and the electrode potential; consequently, the mercury electrode will serve as a pM electrode, provided that the concentrations of the mercury chelonate and the metal chelonate are constant. As these concentrations will be very nearly constant near the end point of a titration of the metal with the chelating agent, the mercury electrode may be used as an indicating electrode.

Other metal electrodes could be used, but the mercury electrode is highly reversible, has a high hydrogen overvoltage and may be used in acidic solutions. In addition, mercury generally forms very stable chelates, and hence most metals will not completely displace mercury from its chelonate; this is a necessary requirement for operation of the electrode because, if all the mercury chelonate is removed from solution, equation (68) will no longer apply and the electrode will no longer function as a pM indicator. The relatively positive standard potentials of the mercury couples minimise the interference of oxidising agents, particularly oxygen from the air. It is true that, for some titrations[220], such as that of barium in ammoniacal solution, the mercury electrode cannot be used in solutions saturated with air. Halides interfere in all circumstances, because of the formation of insoluble mercurous halides, but concentrations as low as $10^{-6}M$ cause no difficulties in practice. To avoid the presence of halides, the potential of the indicator electrode is generally measured against a reference electrode that is connected to the solution through a salt bridge.

Reilley *et al.*[220, 222, 223] have published a number of potential/pH diagrams that can be used to predict the titration curves of metal ions, alone or in admixture with various chelating agents: Fig. III, 23 shows a simplified version of such a diagram for the EDTA system; any tendency for the buffer solutions to form complexes with the various metal ions is neglected. Curve I shows the calculated values of the potential at different pH values for the reaction $Hg + 2OH^- = HgO + H_2O + 2e$, and Curve II the potential/pH function for a solution containing equal amounts of free EDTA and mercury-EDTA complex. The horizontal lines for each metal ion may be measured or calculated from the known stability constants[223].

If a solution containing manganese and calcium ions and having a pH of 6.0 is titrated with EDTA by using the Reilley indicator electrode, the potential measured at the start of the titration will be given by A in Fig. III, 23; as soon as some EDTA had been added and the manganese-EDTA complex has formed, the potential will move to B, and will remain close to B until all the manganese has reacted with the EDTA. Further additions of EDTA cause the potential to jump to point C, and the manganese end point will thus be established. If more EDTA is added, it will react with the calcium, and when all the calcium been complexed, the potential will shift to D, indicating

References p. 165

the position of the second end point. This picture is somewhat oversimplified, and Reilley's work should be consulted for a more detailed discussion.

Some of the metal ions that may be titrated are given in Table III, 5 together with the buffer solutions used. A generally applicable procedure,

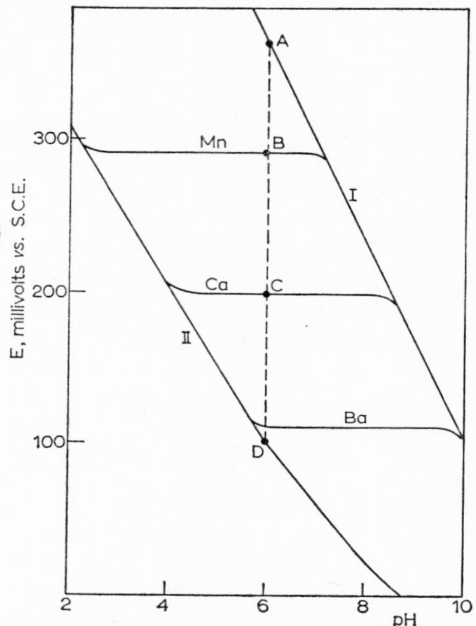

Fig. III, 23. Potential-pH diagram for the EDTA system.

which is a condensed version of the procedures of Reilley and Schmid[221], is given below.

Potentiometric Titration of Metal Ions with EDTA

Apparatus:

Amalgamate a clean gold electrode by dipping it in pure mercury and rinsing it with water, and connect it and a saturated calomel electrode to a potential-measuring device (see Note 1). Connect the calomel electrode with the solution by means of an ammonium nitrate salt bridge.

Reagents:

Ammonium Buffer. Add ammonium hydroxide to a 0.5*M* solution of ammonium nitrate until the pH is 9.5 to 10.

Acetate Buffer. Add acetic acid to a 0.5*M* solution of sodium acetate until the pH is about 4.6.

Mercury(II)-EDTA complex. Mix equivalent amounts of mercuric nitrate and EDTA solutions, and dilute the mixture until the final concentration is 10^{-2} to $10^{-3}M$.

Standard EDTA solution. Weigh the appropriate amount of the pure disodium salt of EDTA, and dissolve it in distilled and de-ionised water.

Procedure:

To a 15 to 25 ml aliquot of the sample solution of the metal ion(s) to be titrated add 25 ml of the appropriate buffer (see Table III, 5) and 1 drop of mercury-EDTA reagent (see Note 2); if the titration is to be carried out at pH 1.5 to 2 add ammonium hydroxide or nitric acid until this pH is attained. Titrate the solution with the standard EDTA solution and take the end point as the point of maximum potential change.

For back-titrations, *e.g.* in the determination of aluminium, heat the solution and add an excess of standard EDTA solution just before the addition of buffer (see Note 3); then add the buffer solution and the mercury-EDTA reagent and titrate with a standard metal ion solution.

Notes:

(1) A potentiometer or a pH meter of high sensitivity may be used.

(2) For some titrations in the ammonium buffer, especially that of barium, oxygen must be removed by the passage of nitrogen at this point.

(3) Some metal ions, such as aluminium and chromium, form EDTA complexes only slowly at room temperature; they must therefore be heated in the presence of an excess of EDTA before complex formation can occur. The excess of EDTA is then back-titrated at room temperature with a metal ion whose complexes form rapidly at room temperature.

Mixtures of metal ions in different pH groups in Table III, 5 may be titrated: for instance, Reilley and Schmid[221], titrated bismuth, cadmium and calcium in one solution. The pH of the mixture was first adjusted to 1.2–2 and the above procedure was then followed; the first potential break corresponded to the amount of bismuth present. The pH was changed to 4 by adding sodium acetate/acetic acid buffer and the titration was continued until a second potential break was obtained; the amount of EDTA used between the first and the second break corresponded to the amount of cadmium. Finally, the pH was adjusted to 8 with ammonium buffer and the calcium was titrated. Accuracies within 2% were obtained for these component mixtures, whereas the results of most single titrations were accurate within 0.4%.

Most metal ions listed under pH 4 to 5.5, or pH 8 to 10, will not interfere with titrations at pH 2; the alkaline earths do not interfere with titrations at pH 4 to 5.5.

References p. 165

TABLE III, 5

CONDITIONS FOR POTENTIOMETRIC EDTA TITRATIONS

pH	*Buffer*	*Metals that may be titrated*
2	Nitric Acid	*DIRECT*: Th, Hg, Bi
4–5.5	Acetic Acid	*DIRECT*: Sc, Y, La, Rare Earths (3+ oxidation state), VO^{2+}, Mn^{2+}, Cu, Zn, Cd, Hg, Bi
		BACK-TITRATION (with Cu, Zn or Pb): Cr^{3+} (heat near boiling for 10 min with excess of EDTA before back-titration) Fe^{3+}, Ni, Al, Ga, In, Zr, Hf, Th, Sc, Y, La, Rare Earths (3+), VO^{2+}, Mn^{2+}, Cu, Zn, Cd, Hg, Bi, Pb
8–10	Ammonium	*DIRECT*: Mg, Ca, Sr (pH 10), Ba (pH 10, remove O_2), Co, Ni, Cu, Zn, Cd, In, Pb (tartrate added to keep Pb and In in solution)
		BACK-TITRATION (with Cu, Zn, Ca, Mg, or Cd): Cr^{3+} (boil for 10 min with excess of EDTA before back-titration), Sc, Y, La, Rare Earths (3+), Hg, Bi, Mg, Ca, Sr (pH 10), Ba (pH 10, remove O_2), Co, Ni, Cu, Cd, Pb, Tl^{3+}

As well as with EDTA, titrations with tetraethylenepentamine[222] and ethylene glycol bis-(β-aminoethyl)-N,N'-tetra-acetic acid[224] have been carried out potentiometrically with a mercury electrode; potential/pH diagrams analogous to that shown in Fig. III, 23 and detailed procedures were given. The ethylene glycol bis-(β-aminoethyl)-N,N'-tetra-acetic acid is useful, because calcium can be titrated in the presence of magnesium. Tetraethylenepentamine is generally more selective than EDTA.

Two polarised mercury (gold coated with mercury) electrodes have been recommended for constant-current potentiometric titrations with EDTA[225]; polarised electrodes have the advantages that they are very simple to make and that their response and stability are very good. If polarised electrode systems are to function, there must be some electro-active species in solution; neither EDTA nor some of the metal ions titrated (for instance calcium) are normally considered electro-active. Thus, another metal ion that is electro-active and whose concentration is related to the concentration of the metal ion being titrated must be added. Martin and Reilley[225] found that the coupling of the mercury electrode with mercury-EDTA gives a system that fulfills the electro-activity requirement. When constant currents of 5 to 10 microamp were passed between the two indicator electrodes, large potential jumps were found at the end points of the titrations of copper or calcium.

In the calcium titration, the solution to be titrated consisted of 30 ml of 0.02*M* mercury-EDTA, 8.5 ml of buffer (2*M* in ammonium hydroxide and 2*M* in ammonium nitrate), 10 ml of sample solution containing the calcium and 120 ml of water. Before the end point, any EDTA formed a complex with calcium, but there was still an excess of calcium present, which tended to displace mercury from its EDTA (Y) complex.

$$Ca^{2+} + Hg(NH_3)Y^{2-} + (x-1)NH_3 = CaY^{2-} + Hg(NH_3)_x^{2+}$$

The following reactions could then occur at the indicator electrodes,

$$\text{cathode: } Hg(NH_3)_x^{2+} + 2e = Hg^0 + xNH_3$$
$$\text{anode: } Hg^0 + xNH_3 = Hg(NH_3)_x^{2+}.$$

These reactions are reversible at a mercury electrode (see current-potential curves in reference 225), so that the potential difference between the electrodes remained small; at the end point, the electrode reactions are

$$\text{cathode: Reduction of } O_2, \quad \text{or (if } O_2 \text{ is removed)}$$
$$NH_4^+ + Hg(NH_3)Y^{2-} + 2e = Hg^0 + 2NH_3 + HY^{3-}.$$
$$\text{anode: } Hg^0 + xNH_3 = Hg(NH_3)_x^{2+} + 2e.$$

The electrode reactions are now quite different, and there is a large potential difference between them to account for the potential jump at the end point. In general, sharper end points are obtained, if the concentration of the buffer added is kept to a minimum to prevent extensive formation of buffer-metal complexes.

The dropping-mercury electrode (DME) has also been used for the detection of the end point of EDTA titrations[226]; the main example investigated was the determination of cadmium in an acetate buffer at pH 4.2: when a cathodic current of 0.5 μA is applied to the DME, a negative shift in potential is found at the end point, owing to the fact that cadmium is reduced more easily than the cadmium-EDTA complex; the size of the potential jump is about 500 mv, but it occurs slightly before the end point; this deviation is corrected by an equation based on theoretical considerations.

Mercury indicator electrodes have also been applied to the detection of the end point of the titration of cyanide[227] with electro-generated mercury (I and II), in a 0.01*M* sodium hydroxide solution. If small amounts of cyanide (~ 0.03 mg) are titrated, the potential break is not very large but, by titrating the solvent to a specific potential near the equivalence-point potential, adding the cyanide sample and titrating to the same potential, results accurate to about 2% can be obtained; this is better than most other methods for the determination of cyanide. Oxygen must be removed, but other interferences were not investigated.

(d) Oxidation-Reduction Reactions

In the great majority of cases, noble-metal electrodes such as platinum or gold are used to indicate the end points of oxidation-reduction titrations and, for normal null-point potentiometry, the indicator electrode potential is usually measured with reference to a calomel electrode; it is sometimes possible to eliminate the reference electrode by using a bimetallic electrode pair such as platinum and tungsten. When a polarised system is desired, two similar platinum electrodes are generally used (see Section 4*c*,*vi*). When strong reductants are used as titrants, more reliable results are often obtained with mercury indicating electrodes, because of the high hydrogen overvoltage on mercury.

Because most redox titrations are followed with noble-metal electrodes, applications in this section will be discussed in terms of the titrant used, instead of the electrode as was done in previous sections. Table III, 6 is provided to allow the reader to see at a glance some of the more important and useful redox titrations. Whenever possible, important conditions are indicated in the table or discussed in more detail in the text; but, where complete experimental conditions cannot be given, literature references are provided.

In some cases, the sample contains the substance to be titrated in the desired oxidation state; very often, however, the titration must be preceded by the appropriate treatment with oxidising or reducing agents, any excess of which should be easy to remove from solution or to render innocuous; many of the procedures considered in the following sections include the necessary pre-treatment. A very useful discussion of reducing and oxidising agents can be found in Kolthoff and Belcher's "Volumetric Analysis"[228], in which there is a particularly good treatment of the use of metal amalgams as reducing agents.

(*i*) *Potassium Permanganate*

One of the most extensively used oxidising titrants is potassium permanganate. A 0.1N solution of potassium permanganate is quite stable, provided that it is free from dust and manganese dioxide; in order to obtain the best results, such a solution should be standardised every week; if more dilute permanganate solutions are used, more frequent standardisation is desirable.

Potassium permanganate solutions are particularly useful, because the oxidation potential of the permanganate/manganous couple is high, and thus allows the determination of a large variety of reducing agents. The Nernst equation for the reduction of permanganate,

$$E = E^0 - \frac{0.0591}{5} \log_{10} \frac{[Mn^{2+}]}{[MnO_4^-]\,[H^+]^8}, \qquad (69)$$

indicates that the oxidising strength of permanganate is dependent on the

TABLE III, 6

POTENTIOMETRIC OXIDATION-REDUCTION TITRATIONS

Reagent	*Substance Determined*	*Remarks*	*References*
Potassium permanganate	Iodide	$0.2-0.5N$ H_2SO_4	229, 230
		Chloride and bromide can be present; CCl_4 used to remove excess of I_2	231, 232
	Iodate, bromate, nitrite, chloride	Iodimetry	233, 234
	Bromide	In presence of HCN	235
	Nitrite	At 45°	236
	Sulphurous acid	$1N$ H_2SO_4, excess MnO_4^-, destroyed with standard iodide, complete titration with MnO_4^-	237, 238
	Oxalate	At 70°	239
	Ferrocyanide	$0.05N$ ferrocyanide $1.5N$ H_2SO_4	240, 241
		Applied to zinc	242
	Ferrous		243, 244, 245
	Uranium	Reduction with Zn amalgam 2% H_2SO_4; 2 breaks	246, 247, 248
	Vanadium	Reduction with Zn amalgam H_2SO_4; 3 breaks	247, 249
	Molybdic acid	Reduction with lead, HCl	249, 250
	Tin		249
	Titanium	Sn reductor	249
	Indigo		249
	Manganous salts	Volhard	251
		Constant current potentiometry; pyrophosphate pH 6–7	259
			256
	Hydrogen peroxide		252, 253
	Platinum		254
	Arsenite	Strongly alkaline	255
	Tellurium		257
	Thallium		258
	Antimony		260
	Lead	Alkaline solution	261
Potassium dichromate	Ferrous		69, 262
	Ferrocyanide	3.5–7% HCl; at 40–65°	263
	Antimony		264
	Tin		264
	Iodide	$1N$ HCl	265
	Hydroquinone		266
	Barium	Indirect	267
	Glycerol		268
	Sulphite		269
	Copper(I)	Indirect through Fe(II)	270
	Carbohydrates		271
	Uranium	Reduced by Pb or Sn(II) Indirect through Fe(II)	272

TABLE III, 6 (*continued*)

Reagent	*Substance Determined*	*Remarks*	*References*
Ceric sulphate	Arsenic	4*N* HCl; ICl catalyst	273
	Chromium	Excess of Ce(IV) back-titrated with oxalate	274
	Ferrocyanide (Zinc)		275, 288
	Hydrogen peroxide		276
	Iodide	Moderately conc. H^+	277
		4–6*M* HCl	278, 291
	Iron		279, 280
	Mercury	Excess of Ce(IV), back-titrated with Fe(II)	281
	Nitrite		282
	Tellurium	Excess of Ce(IV), back-titrated with Fe(II)	281
	Uranium	Reduced by Zn amalgam	283
		titration at 80–90°	284, 285
	Vanadium	at 50–60°	285, 286
	Organic acids and sugars		287
	Copper, Antimony	Reduced by $CrCl_2$	289
	Cobalt	0.5% EDTA and calcium acetate	290
	Platinum		292
Potassium iodate	Iodide	HCl or H_2SO_4,	293, 294, 295
		chloroform present	304
	Sulphite		296
	Thiocyanate	10%HCl, bimetallic polarised electrodes	297
	Antimony		298
	Hydrazine		299
	Arsenic		300
	Tin(II)		301
	Thorium, Lanthanum	Indirect	302
	Ascorbic acid		303
Potassium bromate	Sugars		305
	Iron		305, 306
	Ferrocyanide		306
	Copper	Titanous chloride as a reducing agent	307
	Arsenic, Antimony		308, 309
	Iodide	0.7*N* HCl	310
	Thallium (I)	5% HCl	311
	Aromatic compounds	Substitution with Br_2	312, 313, 316
	Thiocyanate		314
	Tin		315
	Furfural	Molybdate catalyst	317

TABLE III, 6 (*continued*)

Reagent	*Substance Determined*	*Remarks*	*References*
Ferrous sulphate (See potassium dichromate section for back-titrations)	Vanadium		318, 319
		$NaBrO_3$, $K_2S_2O_8$	320
	Chromium	$K_2S_2O_8$	320
		AgO	321
	Manganese	AgO	321
	Cerium	$NaBiO_3$	322
		$K_2S_2O_8$	323
		AgO	321
	Gold	0.2*N* HCl or less	324
	Nitric acid		325
	Nitrates		329
	Vanadate	In uranyl solutions	326
	Chlorite	Alkaline solution	327
	Cobalt	Oxidise with perborate, add excess of ferrous, then excess of dichromate, titrate with ferrous	328
Ferrous-EDTA	Silver		330, 331
	Iron	Ferrous-EDTA complex electro-generated	332
Arsenic(III)	Hypochlorite	$NaHCO_3$ medium	333
	Bromate		308
	Iodine		334
	Dichromate	20% H_2SO_4	335
		1*N* HCl and manganous salts	336
	Lead dioxide		337
Titanium (III)	Ferric iron		338, 339, 348
	Copper	KI at 55°	338, 340
	Antimony	In presence of As at 90°	308
	Uranium	HCl at 55–60°	338, 353
	Mercury (II)		341
	Bismuth	At 80° + tartaric acid	342
	Vanadium		338, 348
	Gold	$KBrO_3$ to oxidise gold	343
		Cl_2 to oxidise gold	324
	Thallium	At 80°, ammonium acetate, ammonium fluoride and chloride	344
	Platinum metals		345, 346, 347 349, 407
	Cerium	Electro-generated Ti(III)	348
	Ferricyanide	Tartrate	350
	Iodate		350
	Molybdate	At 75–80°, 10% HCl	351
	Dyestuffs	Excess of $TiCl_3$, boil, back-titrate, Fe(III)	352
Chromium(II)	Copper	2–15% H_2SO_4 or 5*M* HCl	354, 357
	Dichromate		354
	Iron	2–15% H_2SO_4	17, 354

TABLE III, 6 (*continued*)

Reagent	*Substance Determined*	*Remarks*	*References*
(Chromium(II) *continued*)	Mercury(II)	5% HCl warm	**355**
	Silver	H_2SO_4	**356**
	Gold	2–5% HCl	**356**
	Tin	5*M* HCl at 85°	**357, 358**
	Antimony	5*M* HCl at 85°	**357, 358**
	Vanadium		**359, 361**
	Titanium		**17, 363**
	Molybdate	5*M* HCl at 80–100°	**360, 361, 363**
	Osmium		**362**
	Bismuth		**363**
	Tungsten		**363**
	Selenium, Tellurium	6–9*M* HCl at 70°	**364**
	Anthroquinones nitro, nitroso, and azo acetylenic compounds		**365**
	Perrhenate		**366**
Tin(II)	Iodine		**367, 377**
	Iron(III)	HCl at 75°	**367, 376**
	Dichromate	4% HCl, not accurate	**368, 372**
	Ferricyanide		**368**
	Gold	2% HCl	**368**
	Platinum	2% HCl at 75°	**369**
	Mercury(II)		**370**
	Chromate	40% H_2SO_4	**371, 372**
	Vanadate	At 90–100°	**371**
	Molybdate		**371**
	Titanium		**373**
	Rhenium		**374**
	Phosphorus	Molybdophosphoric acid titrated	**375**
	Bromine		**377**

hydrogen ion concentration. In 1*M* acid, the formal potential *vs*. S.C.E. is about 1.2 v, but drops to only 0.53 v at pH 7. For this reason, and to avoid the possibility of side reactions, standard permanganate solutions are most often used for the titration of acidic solutions.

Iodide ion is easily oxidised to iodine by permanganate, and this titration is accurate, even in the presence of bromide in amounts up to twice that of the iodide[230]. As iodine can be further oxidised to iodate, variable potentials are noticed after the end point, but the break is large enough for the end point to be easily detected. Bromide can also be determined with permanganate in accordance with the following reaction,

$$2MnO_4^- + 11H^+ + 5Br^- + 5HCN = 2Mn^{2+} + 5BrCN + 8H_2O.$$

In this determination, the solution consists of 10 ml of the bromide sample solution, 5 ml of 10% potassium cyanide solution, 10 ml of concentrated sulphuric acid and 100 ml of distilled water; when this solution is titrated with potassium permanganate, a titration curve showing a maximum is obtained[235]. The method may also be applied to mixtures of iodide and bromide.

The titration of ferrous iron[243–245] is very common and very accurate; the potential break occurs just before the appearance of the pink colour due to an excess of permanganate. The reverse titration may also be carried out, provided that the acidity is high and that the titration is carried out rapidly[244, 321]; otherwise, manganese dioxide may be precipitated.

Several determinations of uranium have been recommended[246–248]. The uranium is reduced by passing a mixture of 10 ml of sample and 40 ml of 2% sulphuric acid through a Jones reductor; for the best results, the solution should be warmed to 80–90° and air removed with carbon dioxide before the reduction is carried out. The titration, which is carried out under carbon dioxide, yields one or two potential breaks, according to the conditions[246]. Gustavson and Knudson[247] have titrated mixtures of uranium, vanadium and iron by similar methods; four breaks are obtained in the titration curve, but the results are dependent on the last three only. Between the first and the second break, uranium is oxidised from the 4+ to the 6+ oxidation state and vanadium from 3+ to the 4+; the amount of oxidant used between the second and third breaks is equivalent to the amount of iron. Vanadium(IV) is oxidised to vanadium(V), as indicated by the last potential jump, and all the components can thus be determined, provided that reduction is complete.

The permanganometric oxidations of chromium(II), uranium(III) and (IV), tungsten(IV) and (V), and molybdenum(III) have been studied by Treadwell and Nieriker[378].

Standard permanganate solutions can be used to determine manganese in the 2+ oxidation state[251, 256, 259]. The simplest and most accurate method is that of Lingane and Karplus[256], in which the manganous ion is oxidised to manganic in the presence of a complexing agent, sodium pyrophosphate. The end-point potential and the size of the potential break depend on the pH; at the optimum pH of 6, the potential break is about 300 millivolts. The titration curve should generally be plotted but, if the conditions are reproduced carefully, subsequent determinations may be made by titration to the predetermined end-point potential. In order to increase the ease with which the end point may be found, Huber and Shain[259] have applied constant-current potentiometry with two similar platinum electrodes to the Lingane-Karplus method. (The end point of the classical Volhard manganese titrat-

References p. 165

ion was also found by constant-current potentiometry.) The method of Huber and Shain is described below.

Potentiometric Titration of Manganese

Apparatus:

pH meter for pH adjustment and for the potentiometric titrations; a source of constant current consisting of a 45 volt battery and a 30 megohm resistor (see Note 1). If the constant-current potentiometric method is used, connect two similar platinum electrodes to the pH meter; otherwise, use a platinum electrode and a calomel reference electrode.

Reagents:

Potassium permanganate. 0.1*N* standard solution.
Sodium pyrophosphate. Reagent grade (see Note 2).

Procedure:

Take a sample of such size that the aliquot for titration will contain about 140 mg of manganese (see Note 3) and dissolve it in the appropriate acid. Add an aliquot of the acid sample solution to 200 to 300 ml of saturated sodium pyrophosphate solution contained in a 600 ml beaker, followed by sulphuric acid or sodium hydroxide until the pH is between 5 and 7. Place the indicator electrode(s) in the solution and titrate it with the standard potassium permanganate solution. Plot the titration curve (see Note 4).

Notes:

(1) See section 4*c(vi)*p. 107 for a description of this type of apparatus.

(2) Stock solutions of sodium pyrophosphate tend to be unstable, and should not be prepared until they are wanted.

(3) If this sample size is inconvenient, decrease it by a factor of 10 and use 0.01*N* potassium permanganate.

(4) If the constant-current method is used, take the peak of the titration curve as the end point.

The advantages of the Lingane-Karplus method or its modifications are relative rapidity and almost complete freedom from interference.

(*ii*) *Potassium Dichromate*

Potassium dichromate is a useful oxidimetric titrant, particularly because standard solutions can be prepared directly by weight and are stable for an indefinite period. Dichromate solutions are not such strong oxidants as solutions of permanganate or ceric ions, but they are adequate for many purposes. Although the standard potential of the dichromate/chromic couple is increased by increases in hydrogen ion concentration, the increases are not in agreement with theoretical calculations based on the Nernst equation. There is evidence that chromium in the 5+ oxidation state is an important intermediate in the reduction of dichromate, and the concentration of chromic ions has some effect on the potentials measured but not the theoretical effect; as neither the exact form of the chromium(V) nor its role in the reduction sequence are completely known, ordinary Nernst equation calculations are often in error. It has been noticed that the indicator electrode potential tends to rise (become more oxidising) during the titration of dichromate with ferrous solutions[262]; this non-thermodynamic occurrence may be related to the oxidation of the electrode by chromium(V). Oxidation of the electrode by dichromate is apparently rather slow (page 87).

Dichromate solutions are extensively used for the titration of ferrous solutions, although slight positive errors have been noticed if the solutions are very dilute[69, 262]; this may be due to the slow attainment of potentials at the indicator electrode, again because of the formation of oxide films on the electrode surface.

One of the more interesting determinations with a standard dichromate solution is that of organic substances whose carbon atoms are individually attached to oxygen[271], *e.g.*, carbohydrates or cellulose, and this determination is now given in detail.

Determination of Carbohydrates and other Organic Substances

Apparatus:

pH meter or other potential-measuring device coupled with a platinum and a calomel electrode.

Reagents:

Potassium dichromate. Prepare a 1.835 N solution from an accurately weighed quantity of the solid reagent.

Ferrous ammonium sulphate. 0.5M solution 0.2M in sulphuric acid; standardise this solution frequently against the standard dichromate.

Procedure:

Weigh into a beaker as much of the sample as will react with about 80% of the dichromate and add 25.00 ml of the 1.835N dichromate solution, followed by 10 ml of concentrated sulphuric acid, taking 10 to 15 seconds for the addition. Stir the solution well, avoiding excessive foaming. After 10 minutes, add 150 ml of water and titrate the solution potentiometrically with the standard ferrous solution.

When this procedure is used, the oxidation of the following compounds is at least 99% complete: starch, glucose, fructose, sucrose, arabitol, formic acid, oxalic acid, glycollic acid, methanol, tartaric acid, citric acid, benzoic acid, phthalic acid and a few others. Ethanol, sodium acetate, lactic acid, propionic acid, glycine, acetamide, *etc.* are only incompletely oxidised.

(*iii*) *Ceric Salts*

Various salts of cerium(IV) can be used to prepare oxidising reagents, and because the potential of the ceric/cerous couple is highly oxidising, almost any reducing agent can be titrated; in addition, ceric solutions, at least those in sulphuric acid, are quite stable even when boiled for a few hours.

Extensive information about all the aspects of the use of ceric solutions in oxidation-reduction reactions may be found in the small book [287] prepared by G. F. Smith; in addition to information about the preparation and standardisation of titrant solutions, much information about various determinations reported before 1942 is given.

The most usual method of preparing a ceric solution involves weighing out approximately the required amount of reagent grade ammonium nitratocerate, $(NH_4)_2Ce(NO_3)_6$, and dissolving it in 0.5 — 1*M* sulphuric acid; this "ceric sulphate" solution is then standardised by one of the methods discussed below. The formal potential of the ceric-cerous couple in 1*M* sulphuric acid is 1.44 V *vs.* the normal hydrogen electrode. If a ceric solution of greater oxidising power is needed, it may be made up in perchloric acid or obtained commercially. A wide range of potentials may be selected by the proper choice of acid and its concentration, as is shown in Table III, 7; the potentials in this table are based on the findings of Smith and Getz[379]. Wadsworth, Duke and Goetz[380] later summarised the available information on the complexes of the ceric ion and their effect on the standard potential of the ceric/cerous couple.

TABLE III, 7

FORMAL POTENTIALS OF THE CERIC/CEROUS COUPLE

Normality	$HClO_4$ *volt*	HNO_3 *volt*	H_2SO_4 *volt*	*HCl* *volt*
1	1.70	1.61	1.44	1.28
2	1.71	1.62	1.44	
4	1.75	1.61	1.43	
6	1.82	-	-	
8	1.87	1.56	1.42	

Ceric solutions may be standardised against sodium oxalate [287], arsenious oxide[273, 381] or potassium dichromate (by means of a ferrous solution) among others[287]. Sodium oxalate in a sulphuric acid medium (2.5 ml of H_2SO_4 per 100 ml) may be titrated at 70° with a ceric solution, or the titration may be carried out at room temperature with 5 ml of 0.005*M* ICl as a catalyst; the end point is found potentiometrically. The catalyst may be prepared by dissolving 0.279 g of potassium iodide and 0.178 g of potassium iodate in 250 ml of water. When the solids have completely dissolved, 250 ml of concentrated hydrochloric acid is added, and the equivalence of iodide and iodate is then adjusted potentiometrically by the addition of dilute solutions of one or the other as necessary; where the proper potential adjustment has been obtained, one drop of iodide or iodate solution should cause a large potential change.

Standardisation with arsenious oxide is carried out in 4*M* HCl, also with ICl as catalyst.

Standard ceric solutions have found extensive use in the analysis of steels[286, 287]; for example, Willard and Young[286] have devised the following procedure for the analysis of vanadium in steels which also contain chromium.

Determination of Vanadium in Steels

Apparatus:

pH meter or potentiometer, a platinum indicator electrode, and a calomel reference electrode.

Reagents:

Ceric sulphate. 0.1*N* standard solution (see above).

Ferrous ammonium sulphate. 0.05*N* standard solution.

Procedure:

Transfer 4 to 5 g of a sample of low vanadium (0.2%) steel to a 600 ml beaker, add 40 ml of water and 10 ml of concentrated sulphuric acid. Boil the liquid until salts begin separate, in order to ensure the decomposition of the carbides present. Dissolve the mass in 20 ml of water and slowly add concentrated nitric acid until the violent oxidation of ferrous iron ceases; then boil out the oxides of nitrogen.

Add 2.0 g of ammonium persulphate and 2 ml of a 0.25% solution of silver nitrate, and boil the mixture for fifteen minutes in order to ensure complete oxidation; cool the liquid and add 20 ml of concentrated sulphuric acid.

Place the electrode attached to the potential-measuring device in the solution and add 0.05*N* ferrous solution until an excess of 5 ml has been added as indicated by the potential. After 10 min add a *slight* excess of dilute potassium permanganate, cool the mixture to 5 to 10°, and titrate it with the ferrous solution to the potentiometric end point; or add an excess and back-titrate it at room temperature with standard ceric sulphate solution. Dilute the solution to 300 ml, heat it to 70–75° and titrate the vanadium from the (IV) to the (V) state, using the standard ceric solution.

References p. 165

The oxidation of uranium(IV) to uranium(VI) by ceric ion has been studied[382, 386] and a determination of plutonium has also been devised[383]. The problems encountered in the plutonium determination were considerable, because of the extremely low permissible daily body intake. The plutonium was oxidised to the Pu(VI) state with fuming perchloric acid and then reduced to Pu(IV) with a slight excess of standard ferrous solution. An automatic potentiometric titration with a standard ceric ammonium sulphate solution was then carried out; the method is accurate, but many substances such as chromium, gold, manganese, vanadium and platinum interfere.

Electro-generated ceric ion has been extensively used for the coulometric titration of a number of substances, including iodide[291], iron, titanium, ferrocyanide[384] and uranium[385].

In the titration of chromic ion with standard ceric solutions in the presence of dichromate[387], and ferrous ion [388], the use of a bimetallic electrode pair consisting of a platinum and a tungsten electrode has been studied; in the presence of the ferrous ion, it was found possible to predict the form of the titration curves on the basis of current-potential curves. In oxidising solutions, it was shown that the tungsten electrode became coated with a film of tungstic oxide.

(*iv*) *Potassium Iodate and Bromate*

Potassium iodate and potassium bromate are easily obtained pure, and their standard solutions may be made up from accurately weighed quantities. Generally, the standard iodate or bromate solution is added to the solution of the sample, to which has been added a large amount of the corresponding halide. A reaction then takes place between, for example, bromate and bromide with the production of bromine, ($6H^+ + BrO_3^- + 5Br^- = 3Br_2 + 3H_2O$), and it is the bromine that oxidises the substance being titrated; as bromine is a stronger oxidant than iodine, potassium bromate solutions are more extensively used.

The oxidation of mixtures of tervalent arsenic and tervalent antimony with potassium bromate normally yields only one potential break. Quinquevalent antimony may, however, be selectively reduced in the presence of quinquevalent arsenic by the careful addition of titanous ion[309]. If a 5% solution of hydrochloric acid containing arsenic and antimony is heated and reduced with a titanous solution to a potential of 0.3 volt *vs.* S.C.E., only the antimony will be reduced, provided that the excess of titanous ion is quickly removed; a few drops of a 3% solution of copper sulphate and stirring in air rapidly removes the excess of titanous ion, as is indicated by an increase of the indicator electrode potential to 0.5 V; the antimony is then determined by means of a standard potentiometric titration with potassium bromate.

Potassium bromate solutions can also be used for the bromination of a number of organic compounds[312]; the procedure is described below.

Potentiometric Bromination of Organic Compounds

Apparatus:

pH meter or potentiometer, two indicator platinum electrodes and a salt bridge.

Reagents:

Potassium bromate. 0.05*N* standard solution. Weigh out pure $KBrO_3$ and dissolve it in a known volume of water.

Potassium bromide. 20% (w/v) solution.

Reference solution. 10 ml of concentrated HCl, 10 ml of 20% KBr solution, 1 drop of $KBrO_3$ solution and 80 ml of water.

Procedure:

Dissolve a weighed portion of the sample as completely as possible in 100 ml of a solution containing 15 ml of concentrated hydrochloric acid and 10 ml of 20% potassium bromide solution. Place one electrode in the sample solution and the other in the reference solution, connect the two solutions by means of the salt bridge, and titrate the sample with the standard potassium bromate solution until there is no potential difference between the electrodes. (Pinkhof-Treadwell method.)

This procedure has been applied to aniline, *o*-toluidine, *m*-toluidine, *p*-toluidine, phenol, nitrophenol, and sulphanilic acid[312].

(*v*) *Other Oxidants*

Standard solutions of several other oxidants may be prepared. Hypobromite solutions have been used as oxidants in neutral or alkaline media[389], and standardised against arsenious oxide; this reaction is more accurate in dilute solution. At a concentration of 10^{-3} *N*, an accuracy of $\pm$ 1% of the material determined can be obtained. Hypobromite may be used to determine ammonia according to the following reaction

$$2NH_3 + 3NaOBr = 3NaBr + N_2 + 3H_2O.$$

To obtain good results, the hypobromite is added in excess, which is back-titrated with a standard solution of arsenious oxide.

In basic solutions, potassium ferricyanide is a strong oxidant, and can be used to titrate chromic ions, if thallous ion is added as a catalyst[410]; stannous, arsenious, tervalent antimony and vanadyl solutions may also

References p. 165

be titrated potentiometrically with standard potassium ferricyanide solutions in basic media[391]. Cerous cerium can be titrated in concentrated, air-free solutions of potassium carbonate[392], and so can cobalt and manganese[393, 394].

Solutions of sodium vanadate have been recommended for the potentiometric titration of tervalent titanium and tervalent molybdenum[395].

Sodium biphenyl has been used as a reagent for the potentiometric titration of aromatic hydrocarbons dissolved in tetrahydrofuran[396]: the potential of a platinum indicator electrode was measured against another platinum electrode placed in the tip of the burette; this arrangement eliminates the difficulties often encountered with ordinary reference electrodes in non-aqueous media.

(vi) Tervalent Arsenic

Arsenious oxide can be obtained in a very pure state and is easily weighed out for the preparation of standard solutions; it is usually dissolved in the least possible amount of 1*M* sodium hydroxide, and the solution is then acidified and diluted to the appropriate volume. Alkaline solutions of arsenious oxide are slowly oxidised by air, but neutral or slightly acid solutions are quite stable.

Standard solutions of tervalent arsenic can be used to titrate dichromate in 20% sulphuric acid[335]; vanadate does not interfere as it is not reduced by tervalent arsenic, and it may be titrated with a standard ferrous solution after the determination of the dichromate; low results are obtained in the presence of manganese, because the higher oxidation states of manganese form complexes with arsenic acid, and their higher stability causes induced reduction of the dichromate. Vanadium also can take part in the induced reduction. The following series of reactions has been suggested as an explanation of these phenomena:

2Cr(VI)	+ As(III)	= 2Cr(V)	+ As(V)	fast
Cr(V)	+ As(III)	= Cr(III)	+ As(V)	slow
Cr(V)	+ 2Mn(II)	= Cr(III)	+ 2Mn(III)	very fast
Cr(V)	+ Mn(III)	= Cr(VI)	+ Mn(II)	,, ,,
2Mn(III)	+ As(III)	= 2Mn(II)	+ As(V)	very slow
Cr(V)	+ V(V)	= Cr(VI)	+ V(IV)	
V(IV)	+ Mn(III)	= V(V)	+ Mn(II)	

On the basis of this study, Zintl and Zamis[335] were able to formulate a procedure for the determination of chromium and vanadium in steel: manganese was removed to prevent its interference, but a controlled amount was re-added to catalyse the reaction between the dichromate and the standard solution of tervalent arsenic; the detailed procedure is given below.

Determination of Chromium and Vanadium in Steel

Apparatus:

Potentiometer or pH meter provided with a platinum electrode and a calomel reference electrode.

Reagents:

Arsenious oxide. Standard solution, 0.1*N* or less according to the composition of the sample.

Ferrous ammonium sulphate. Standard solution, 0.1*N* or less according to the composition of the sample; standardise this solution daily against potassium dichromate.

Hydrogen peroxide. 3% (w/v) solution.

Manganese sulphate. Dissolve 120 mg of $MnSO_4 \cdot 7H_2O$ in a litre of water.

Procedure:

Fuse a weighed sample (6–8 g usually suffices) with sodium peroxide in an iron crucible. Dissolve the cooled melt in a beaker containing 100 ml of 2*M* sodium hydroxide; remove the crucible from the beaker and wash it with the least possible amount of water.

Add a few ml of hydrogen peroxide, boil the solution for about 10 minutes until it changes from green (manganate) or red (ferrate) to pure yellow, filter it through a sintered glass crucible and wash it with 2*M* sodium hydroxide. Transfer the filtrate and washings to a volumetric flask and dilute the solution to the mark.

Transfer an appropriate aliquot to the titration beaker, slowly add 100 ml of cold 1 : 1 sulphuric acid followed by 2–3 ml of the manganese sulphate solution, and titrate the solution potentiometrically with the standard arsenious oxide solution until the chromium is reduced.

If tungsten is present, add 30 ml of phosphoric acid. Titrate the vanadium with standard ferrous solution.

The combination of manganese and iodide has also been recommended as a catalyst for the titration of dichromate with solutions of arsenious oxide[336].

(*vii*) *Bivalent Iron*

Solutions of ferrous ammonium sulphate have found extensive use in potentiometric redox titrimetry. Although the ferrous ion is not a very strong reducing agent, ferrous solutions are not very stable, and their stability is largely dependent on their acidity; in order to ensure maximum stability, solutions of ferrous ammonium sulphate are usually made up in about 0.1*N* sulphuric acid. If carefully stored to prevent loss of water of hydration, the reagent can be used as a primary standard; but, as ferrous

References p. 165

solutions are not stable, they are usually made up roughly and standardised daily against potassium dichromate.

Ferrous solutions are often used in steel analysis for the titration of vanadate[318, 319] or dichromate[320, 321, 397]. The vanadium, chromium or manganese present must first be oxidised, and many oxidising agents have been suggested, among them being potassium persulphate and sodium bismuthate; the excess of such reagents must be removed by boiling or filtration before the titration is carried out. A more convenient oxidising agent is argentic oxide[321], any excess of which may be decomposed by gentle warming or titration with a ferrous solution. (The first potential break indicates that the reduction of the argentic oxide is complete, and the second can be used to calculate the amount of manganese or chromium.) The method is described below.

Determination of Manganese, Cerium or Chromium

Apparatus:

Standard potentiometric apparatus equipped with a platinum electrode.

Reagents:

Argentic oxide, AgO (Merck and Co., Rahway, N.J., U.S.A.).

Ferrous ammonium sulphate. 0.1*N* solution. Dissolve the appropriate, accurately weighed, amount of ferrous ammonium sulphate in 0.1*N* sulphuric acid.

Procedure:

Add enough nitric acid and water to the sample solution to give a final volume of about 50 ml and a nitric acid concentration of 2–5*M*. Add, successively, small portions of solid argentic oxide until an excess is present as indicated by the dark brown or black colour of the solution. Dilute the solution to about 150 ml with 1*M* sulphuric acid, warm it to about 80° and allow it to stand until the silver oxide has decomposed; this takes about 15 minutes (see Note). Titrate the solution potentiometrically with standard ferrous ammonium sulphate solution.

Note:

Alternatively, the titration may be started without heating or waiting; the amount of oxidising substance present is then calculated from the volume of ferrous solution used between the first and second potential breaks.

If this procedure is applied to steel samples, the sum of the manganese, vanadium, and chromium is determined. If the sum of the manganese and chromium is required, the vanadium can be determined by adding an excess of ferrous solution and back-titrating with a ceric solution: the vanadium will be reduced by the ferrous solution and re-oxidised by the ceric solution, the volume of which gives a measure of the vanadium. Manganese must be removed previously if the amount of chromium is required.

The reducing power of ferrous solutions may be increased by the addition of a reagent that strongly complexes the ferric ions: for instance, it is common practice to add phosphoric acid to ferrous solutions that are to be titrated with dichromate, as this increases the sharpness of the end point. If a very strong complexing agent such as EDTA is used, the ferrous ion becomes a very powerful reductant; silver[330, 331], copper[398], iodine[399] and ferric iron[332] have all been determined with this reagent. The determination of ferric iron was carried out by potentiometric titration with the electro-generated ferrous-EDTA complex formed by the coulometric reduction of the ferric-EDTA complex; the conditions had to be carefully controlled and the titration could not be used for other oxidants owing to slow reactions or interferences.

(*viii*) *Titanous Ion*

The titanous ion is a very strong reducing agent, and can be used for the potentiometric titration of a wide variety of oxidising agents. The standard potential for the reaction

$$TiO^{2+} + 2H^+ + e = Ti^{3+} + H_2O$$

is +0.1 V *vs.* N.H.E., but this varies with the composition of the solution in which the titration is carried out.

Titanous chloride and sulphate are available commercially, often in solution. Many commercial materials contain significant amounts of iron, which is undesirable for some applications; the presence of iron can easily be detected by titrating the suspected solution with potassium dichromate and recording the potentiometric titration curve.

A convenient and inexpensive way of preparing iron-free titanous solutions from titanium hydride has been suggested[400, 401], and is described below.

Preparation of Titanous Chloride from Titanium Hydride

Reagent:

Titanium hydride (Metal Hydrides, Inc., Beverly, Mass., U.S.A.).

Procedure:

For 5 litres of approximately 0.4*N* titanous chloride, place 55 g of titanium hydride (a 10% excess) in each of two 2-litre flasks, and add 500 ml of 37% hydrochloric acid to each. Heat the flasks gently until the reaction starts, and then cool or heat each flask to maintain it at the desired rate until solution is complete. Digest the solution overnight on a steam bath, and filter it through

References p. 165

a pad of glass wool into 4 litres of air-free water (prepared by bubbling nitrogen or carbon dioxide through it). If possible, allow the solution stand for 10 days before use.

The standardisation of titanous solutions may be effected in a number of ways, which have been critically reviewed by Lamond[402]. If the titanous solution is free from iron, the simplest method seems to be the titration of potassium dichromate in 10% sulphuric acid[403]; if the standard solution contains iron, the effect of this on the determination to be carried out must be considered. The standardisation of titanous solutions that contain iron may be effected with dichromate as follows: an excess of ferrous iron is added to the dichromate solution, which oxidises its own equivalent to the ferric state; the ferric iron is then quantitatively reduced with the titanous solution[402].

Standardisation may be carried out in hydrochloric acid solution[404], and potassium iodate or bromate may be used as primary standards[350].

Because titanous solutions are easily oxidised by air, they must be stored under an inert atmosphere, and the solutions being analysed must also be free from air and titrated in an inert atmosphere; titanous solutions should not be allowed to come into contact with rubber. If these precautions are observed, the standardisation of the solution need only be repeated at two-week intervals.

Titration apparatus based on that of Thornton and Chapman[422], and illustrated in Fig. III, 24, keeps the standard solution out of contact with air and allows titrations to be carried out in an inert atmosphere, even at elevated temperatures. Stirring is conveniently effected by the stream of carbon dioxide, or some other inert gas, used to prevent air from dissolving in the solution being titrated.

Many of the disadvantages of titanous solutions can be eliminated by the use of electro-generated titanous ions[348, 405, 406], but air must still be removed from the solution being titrated.

Despite its strong reducing power, many of the reactions of the titanous ion are rather slow for titrations; for this reason, it is often necessary to work at elevated temperatures, or to add an "accelerator" or catalyst. The reduction of ferric iron[338], for example, is so slow at room temperature that titrations must be carried out at 50–60°. It is also possible to speed up the reaction by the addition of 7 mg of bismuth per 100 ml, provided that the concentration of hydrochloric acid is not greater than 10%. Iron may be determined in the presence of copper, if 0.5 g of potassium thiocyanate is added to precipitate the copper[338].

Uranium can be determined with titanous ion[338, 353] between 50° and 60°; tartrate is added to the 1–10% hydrochloric acid, in which the titration is carried out, in order to keep the quadrivalent titanium in solution. Ferrous ion may be used as a catalyst for the determination of uranium[405].

The titration of mercuric ions with titanous solutions yields metallic mercury, provided that some bismuth salt is added[341]. The titration is carried out in a solution of 2 g of tartaric acid, 3–10 g of ammonium acetate (the amount depends on the acidity), 15 g of ammonium chloride, 5–10 ml of

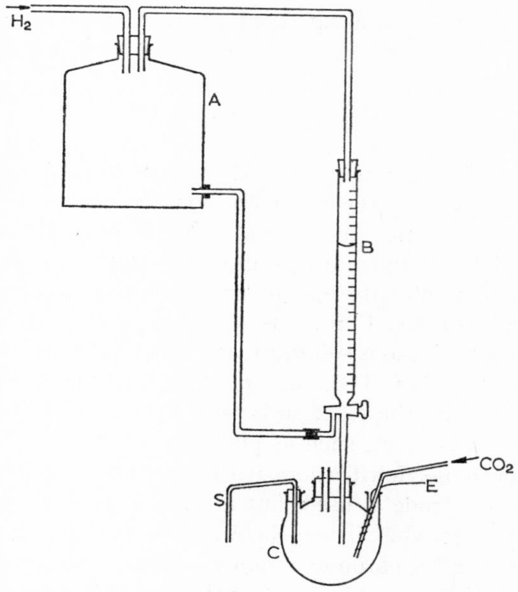

Fig. III, 24. Titration apparatus for use with strong reductants.

0.1*N* bismuth chloride and 200 ml of water; air must be removed with an inert gas. Bismuth may also be titrated in this medium. Nitrates interfere.

Methods have been devised for the determination of platinum metals by potentiometric titration with titanous solutions. Platinum, palladium, ruthenium, rhodium, and iridium[345] have all been titrated with the addition of citrate to increase the potential break at the end point. Titrations at 70° of these metals, as well as of gold and osmium[346], have been investigated. Müller and Stein[407] have titrated platinum after oxidation by chlorine, and found two potential breaks in titrations in hydrochloric acid at 60°; the amount of titrant consumed between the first and second breaks was equivalent to the reduction of quadrivalent platinum to the bivalent state. Effec-

References p. 165

tive stirring and slow addition of the titrant were necessary to prevent the precipitation of platinum metal.

Osmium was best titrated in a 0.3*N* solution of hydrobromic acid[347]. The titration proceeded rapidly at 60°, but the temperature could not be raised until near the end point, because octovalent osmium is volatile; for this reason, the removal of air also proved troublesome.

A number of dyestuffs may be determined by adding excess of titanous chloride, boiling for five minutes, and back-titrating with a standard ferric solution[352]. Many other oxidising agents, such as permanganate, quinquevalent antimony, tervalent thallium, chlorate, and dichromate[408], have been titrated.

(ix) Chromous Ion

The most powerful reductant available to analytical chemists in the form of a standard solution is the chromous ion. Chromous solutions may be prepared by the reduction of an acid solution of potassium dichromate with amalgamated zinc, and may be standardised against potassium dichromate[354, 409], or copper sulphate[354], or even prepared from accurately weighed quantities of dichromate. Chromous solutions should be stored with the same precautions as titanous solutions, and the apparatus shown in Fig. III, 24 can be used for either. The chromous ion is powerful enough to reduce hydrogen ions, though the reaction is very slow, except in the presence of finely divided heavy metals such as platinum.

In most potentiometric titrations with chromous solutions, a bright platinum indicator electrode is used, but a larger potential break is obtained with a mercury electrode[17]. This is because the mercury electrode indicates the true potential of the chromic/chromous couple. The potential of the platinum electrode is influenced by the fact that the reduction of hydrogen ions by the chromous ion is catalysed at the platinum surface, causing the actual ratio $[Cr^{2+}]:[Cr^{3+}]$ to be lower at the electrode surface than in the body of the solution.

The procedure for the preparation of standard chromous solutions described by Lingane and Pecsok[410] is given below.

Preparation of Standard Chromous Solutions

Apparatus:

The storage and titration apparatus shown in Fig. III, 24, and a potentiometer.

Reagents:

Amalgamated zinc. Stir pure granulated zinc in a mercuric chloride solution in dilute hydrochloric acid and wash the amalgam with pure water.

Procedure:

Preparation. Fill the storage bottle in Fig. III, 24 about half-full with the amalgamated zinc.

Dissolve exactly 29.421 g of pure, dry potassium dichromate in 500 ml of water in a 2-litre flask and acidify the solution with 27.8 ml of concentrated sulphuric acid. Slowly add, in small portions, 75 ml of 30% (w/v) hydrogen peroxide, which reduces the dichromate to chromic ion. Heat the solution until the evolution of oxygen ceases; this takes about 20 min. Cool the solution, transfer it to a 2-litre volumetric flask and dilute it to the mark. Rinse out the storage bottle with two 100 ml portions of the solution and fill it with the remainder. Close the flask and connect it to the source of hydrogen, *e.g.* a Kipp generator. Complete reduction to the chromous state takes from 30 minutes to a few hours, depending on the amount of zinc present. When reduction is complete, the solution will be a pale blue with no trace of green (see Note 1).

Solutions of chromous chloride in 0.1*N* hydrochloric acid may be prepared in the same way by using the appropriate volume of hydrochloric acid instead of sulphuric acid.

Standardisation (see Note 2). Place an aliquot of pure standard copper sulphate solution in the titration flask (Fig. III, 24) and add 100 ml of 5*M* hydrochloric acid. Displace the air with carbon dioxide or some other inert gas and titrate the solution potentiometrically with the chromous solution.

Notes:

(1) Solutions prepared in this way are stable for 1 to 2 weeks if the acidity is 0.1*N*, or for 3 weeks if the acidity is 1*N*. The period of stability depends on the speed at which hydrogen ion is reduced by the amalgamated zinc.

(2) As the chromous solution was prepared from a weighed amount of dichromate, standardisation is not necessary, unless some doubt exists about the concentration.

Standard chromous solutions may be used for the determination of mixtures of vanadium and iron[359] in 15% sulphuric acid. Two potential breaks are obtained, the first for the reduction of iron and the second for the reduction of quadrivalent vanadium to the tervalent state. Mixtures of iron and titanium[17] and of antimony, copper and tin[357] can also be determined; a mercury indicator electrode was used in the first case, and 5*M* hydrochloric acid was found to be a useful titration medium in the second. The simultaneous determination of selenium and tellurium has also been studied[364].

Nitrate may also be determined by the use of chromous solutions[411] according to the reaction

$$8Cr^{2+} + NO_3^- + 10H^+ = 8Cr^{3+} + NH_4^+ + 3H_2O.$$

Although this reaction is too slow for a direct potentiometric titration to be

References p. 165

feasible, a convenient procedure was devised by using titanic ion as a catalyst, and ferric ion to oxidise the excess of chromous ion. The details of the method are as follows.

Determination of Nitrate Ion

Apparatus:

The apparatus shown in Fig. III, 24, and a potentiometer.

Procedure:

Add about 5 ml of an approximately 0.1*N* solution of titanic sulphate to a 25 ml sample of nitrate in the titration flask and adjust the acidity to 0.5 to 1*N* with sulphuric acid. Heat the solution to $50 \pm 3°$ and remove the dissolved oxygen by the passage of an inert gas. Add rapidly from a burette a 10 to 100% excess of chromous sulphate solution. After 3 to 5 minutes, add an excess of air-free standard ferric solution, insert the platinum electrode and salt bridge and titrate the excess of ferric ion potentiometrically with the standard chromous sulphate solution; the equivalence-point potential is +0.10 V *vs.* S.C.E.

Carry out a blank determination on the same volume of standard ferric ion solution and subtract the blank value from the total volume of standard chromous sulphate used; the difference gives the volume of chromous solution equivalent to the nitrate ion.

(*x*) *Other Reductants*

Solutions of stannous ion in acidic bromide or chloride solution are easy to prepare but, for stability, they must be stored in an air-free atmosphere. Although stannous solutions are not such strong reductants as chromous or titanous solutions, they have been used for a number of determinations, as shown in Table III, 6, p.145. Stannous ions may also be generated electrically for use in coulometric titrations[377, 412, 413].

Standard solutions of vanadous ions have been used for the titration of copper[414], iron(III)[414], silver[414], dichromate[414], cerium(IV)[415], titanium[117], and organic compounds[417] including anthraquinone[416].

Mercurous nitrate solutions have been recommended for the titration of permanganate[418]. The manganese in a steel sample is oxidised with sodium bismuthate and then titrated under carefully controlled conditions; the titration reaction is rather unusual:

$$4MnO_4^- + 7Hg_2^{2+} + 20H^+ = 3MnO_2 + Mn^{2+} + 14Hg^{2+} + 10H_2O.$$

Chromium and manganese do not interfere. Mercurous nitrate in the presence of thiocyanate has also been recommended as a titrant[419].

The use of quadrivalent uranium has been investigated[420], but was found to be not particularly convenient, although it was not oxidised very rapidly by air.

Standard ascorbic acid solutions can be used for the potentiometric titration of gold(III)[421]. The best results are obtained if the titration is carried out at 50° in a solution of pH 1.6 to 3. Chloride up to 0.1*M* can be tolerated, and there is no interference by mercury(II), copper(II) or iron(III).

References

1. W. M. Latimer, *Oxidation States of the Elements and Their Potentials in Aqueous Solutions* (Prentice-Hall, Inc., Englewood Cliffs, N.J., 1952), pp. 340–348.
2. J. J. Lingane, *Electroanalytical Chemistry* (Interscience Publishing Co., New York, 2nd Edition, 1958), Chap. 3.
3. I. M. Kolthoff and N. H. Furman, *Potentiometric Titrations* (John Wiley and Sons, Inc., New York, 2nd Edition, 1931), Chap. 2.
4. P. S. Roller, *J. Amer. Chem. Soc.*, 1928, **50**, 1.
5. I. M. Kolthoff and E. J. A. Verzijil, *Rec. Trav. chim.*, 1923, **42**, 1056.
6. R. W. Schmid and C. N. Reilley, *J. Amer. Chem. Soc.*, 1956, **78**, 5513.
7. R. Přibil, Z. Koudela, and B. Matyska, *Coll. Czech. Chem. Comm.*, 1951, **16**, 80.
8. E. R. Nightingale, *Analyt. Chem.*, 1958, **30**, 267.
9. E. H. Swift, *A System of Chemical Analysis* (W. H. Freeman and Co., San Francisco, 1938), pp. 540–543.
10. I. M. Kolthoff and N. H. Furman, *op. cit.*, pp. 66–67.
11. I. M. Kolthoff and N. Tanaka, *Analyt. Chem.*, 1954, **26**, 632.
12. J. W. Ross and I. Shain, *Analyt. Chem.*, 1956, **28**, 548.
13. J. K. Lee, R. N. Adams and C. E. Bricker, *Analyt. Chim. Acta*, 1957, **21**, 321.
14. F. Anson and J. J. Lingane, *J. Amer. Chem. Soc.*, 1957, **79**, 4961.
15. I. M. Kolthoff and E. R. Nightingale, *Analyt. Chim. Acta*, 1957, **17**, 329.
16. D. G. Davis and J. J. Lingane, *Analyt. Chim. Acta*, 1958, **18**, 245.
17. J. J. Lingane, *Analyt. Chem.*, 1948, **20**, 797.
18. J. J. Lingane, *Electroanalytical Chemistry* (Interscience Publishing Co., New York, 2nd Edition, 1958), p. 23.
19. R. G. Bates, *Electrometric pH Determinations* (John Wiley and Sons, New York, 1954).
20. *Laboratory Practice*, 1959, **8**, 11.
21. H. A. Robinson, *Trans. Electrochem. Soc.*, 1947 preprint **92–38**, 503.
22. H. Ziegal, *Trans. Amer. Electrochem. Soc.*, 1914, **26**, 91.
23. W. E. Shenk and F. Fenwick, *Ind. Eng. Chem., Analyt.*, 1933, **5**, 194.
24. J. J. Lingane, *Analyt. Chem.*, 1948, **20**, 285, 797.
25. *Idem, U.S. Patent* 2,650,256 (1943).
26. R. C. Hawes, A. Stricker, and T. H. Peterson, *Electrical Manufacturing*, May 1951, p. 76.
27. H. H. Baker and R. H. Müller, *Trans. Electrochem. Soc.*, 1939, **76**, 75.
28. H. V. Malmstadt and E. R. Fett, *Analyt. Chem.*, 1954, **26**, 1348; 1955, **27**, 1757.
29. E. H. Sargent and Co., *Scientific Apparatus and Methods*, 1955, **7**, 2.
30. H. V. Malmstadt and D. A. Vassalo, *Analyt. Chem.*, 1959, **31**, 862.
31. H. V. Malmstadt and C. B. Roberts, *Analyt. Chem.*, 1956 **28**, 1884.
32. I. M. Kolthoff and N. H. Furman, *Potentiometric Titrations* (John Wiley and Sons Inc., New York, 2nd. Edition 1931), Chap. 2.
33. J. C. Hostetter and H. S. Roberts, *J. Amer. Chem. Soc.*, 1919, **41**, 1337.
34. E. Müller, *Elektrometrische Massanalyse* (T. Steinkopff, Dresden, 6. Auflage, 1942).

35. W. D. Cooke, C. N. Reilley, and N. H. Furman, *Analyt. Chem.*, 1951, **23**, 1662.
36. J. Pinkhof, *Dissertation*, Amsterdam, 1919.
37. W. D. Treadwell and L. Weiss, *Helv. Chim. Acta*, 1919, **2**, 680.
38. R. Lang, *Z. Elektrochem.*, 1926, **32**, 454.
39. D. C. Cox, *J. Amer. Chem. Soc.*, 1925, **47**, 2138.
40. D. A. MacInnes and P. T. Jones, *J. Amer. Chem. Soc.*, 1926, **48**, 2831.
41. D. A. MacInnes and M. Dole, *J. Amer. Chem. Soc.*, 1929, **51**, 1119.
42. D. A. MacInnes and I. A. Cowperthwaite, *J. Amer. Chem. Soc.*, 1931, **53**, 555.
43. I. M. Kolthoff and N. H. Furman, *Potentiometric Titrations* (John Wiley and Sons Inc., New York, 2nd Edition, 1931), pp. 112–122.
44. I. M. Kolthoff, *Analyt. Chem.*, 1954, **26**, 1689.
45. H. H. Willard and F. Fenwick, *J. Amer. Chem. Soc.*, 1922, **44**, 2504.
46. G. Van Name and F. Fenwick, *J. Amer. Chem. Soc.*, 1925, **47**, 9.
47. N. H. Furman, *J. Amer. Chem. Soc.*, 1928, **50**, 273; 277.
48. E. Müller and H. Kogert, *Z. phys. Chem.*, 1928, **136**, 437.
49. B. Kamienski, *Z. phys. Chem.*, 1929, **145**, 48.
50. J. C. Brünnich, *Ind. Eng. Chem.*, 1925, **17**, 631.
51. R. M. Fuoss, *Ind. Eng. Chem. Analyt.*, 1929, **1**, 125.
52. M. L. Holt and L. Kahlenberg, *Trans. Amer. Electrochem. Soc.*, 1929, **56**, 201.
53. N. H. Furman, *Ind. Eng. Chem. Analyt.*, 1942, **14**, 372.
54. C. N. Reilley, W. D. Cooke and N. H. Furman, *Analyt. Chem.*, 1951, **23**, 1223.
55. G. Duyckaerts, *Analyt. Chim. Acta*, 1953, **8**, 57.
56. C. Bertin, *Analyt. Chim. Acta*, 1951, **5**, 1.
57. G. Charlot and D. Bezíer, *Méthodes Electrochimiques D'Analyse* (Masson et Cie, Paris, 1954).
58. E. Bishop, *Mikrochim. Acta*, 1956, 619.
59. E. Bishop, *Analyst*, 1958, **83**, 212.
60. R. Gauguin, *Analyt. Chim. Acta*, 1958, **18**, 29.
61. I. M. Kolthoff and N. H. Furman, *Potentiometric Titrations* (John Wiley and Sons Inc., New York, 2nd Edition, 1931).
62. N. H. Furman, *Ind. Eng. Chem. Analyt.*, 1942, **14**, 367.
63. *Idem*, *Analyt. Chem.*, 1950, **22**, 33.
64. *Idem*, *ibid.*, 1951, **23**, 21.
65. *Idem*, *ibid.*, 1954, **26**, 84.
66. C. N. Reilley, *Analyt. Chem.*, 1956, **28**, 671.
67. *Idem*, *ibid.*, 1958, **30**, 765.
68. J. T. Wood, H. J. Sand, and D. J. Law, *J. Soc. Chem. Ind.*, 1911, **30**, 871.
69. J. H. Hildebrand, *J. Amer. Chem. Soc.*, 1913, **35**, 847.
70. I. M. Kolthoff and N. H. Furman, *Potentiometric Titrations* (John Wiley and Sons Inc., New York, 2nd Edition, 1931), p. 204.
71. R. E. Cornish, *J. Amer. Chem. Soc.*, 1928, **50**, 3310.
72. W. D. Treadwell and L. Weiss, *Helv. Chim. Acta*, 1920, **3**, 433.
73. I. M. Kolthoff and T. Kameda, *J. Amer. Chem. Soc.*, 1929, **51**, 2888.
74. J. P. Schwing and L. B. Rogers, *Analyt. Chim. Acta*, 1956, **15**, 379.
75. J. T. Stock, W. C. Purdy and Th. R. Williams, *Analyt. Chim. Acta*, 1959, **20**, 73.
76. W. Böttger, *Z. phys. Chem.*, 1899, **24**, 253.
77. I. M. Kolthoff, *Rec. Trav. chim.*, 1922, **41**, 787.
78. M. L. Malprade, *Ann. Chim. (France)*, (10), 1929, **11**, 104.
79. G. L. Wendt and A. H. Clarke, *J. Amer. Chem. Soc.*, 1923, **45**, 881.
80. M. G. Mellon and V. M. Morris, *Ind. Eng. Chem.*, 1924, **16**, 123; 1925, **17**, 145.
81. H. T. S. Britton, *J. Chem. Soc.*, 1925, **127**, 2110, 2142, 2796, 2956.
82. G. A. Perley, *Trans. Electrochem. Soc.*, 1948, **92**, 485.
83. G. Tammann and F. Renege, *Z. anorg. Chem.*, 1926, **156**, 85.
84. N. H. Furman, *J. Amer. Chem. Soc.*, 1922, **14**, 600.
85. G. H. Montillon and N. S. Cassel, *Trans. Amer. Electrochem. Soc.*, 1924, **25**, 259.

86. W. D. Treadwell and G. Schwarzenbach, *Helv. Chim. Acta*, 1928, **11**, 386.
87. W. T. Richards, *J. Phys. Chem.*, 1928, **32**, 990.
88. G. A. Harlow, C. M. Nobel and G. E. A. Wyld, *Analyt. Chem.*, 1956, **28**, 784.
89. I. Shain and G. R. Svoboda, *Analyt. Chem.*, 1959, **31**, 1857.
90. K. W. Franke and J. J. Willaman, *Ind. Eng. Chem.*, 1928, **20**, 87.
91. E. J. Roberts and F. Fenwick, *J. Amer. Chem. Soc.*, 1928, **50**, 2125.
92. H. T. S. Britton, *Hydrogen Ions* (Van Nostrand, Princeton. N.J., 4th Edition, 1956), Vol I.
93. H. T. S. Britton and R. A. Robinson, *J. Chem. Soc.*, 1931, 458.
94. I. M. Kolthoff and B. D. Hartong, *Rec. Trav. chim.*, 1925, **44**, 113.
95. J. J. Shukoff and V. M. Gortikoff, *Z. Elektrochem.*, 1929, **35**, 853.
96. W. D. Treadwell and E. Bernasconi, *Helv. Chim. Acta*, 1930, **13**, 500.
97. H. T. S. Britton and E. N. Dodd, *J. Chem. Soc.*, 1931, 829.
98. L. Kahlenberg and A. C. Krueger, *Trans. Amer. Electrochem. Soc.*, 1929, **56**, 201.
99. A. H. Wright and F. H. Gibson, *Ind. Eng. Chem.*, 1927, **19**, 749.
100. J. C. Brünnich, *Ind. Eng. Chem.*, 1925, **17**, 631.
101. T. P. Hoar, *Proc. Roy. Soc.*, 1933, **A142**, 628.
102. H. C. Parker, *Ind. Eng. Chem.*, 1925, **17**, 737.
103. E. Bülmann, *Ann. Chim.*, 1921, **15**, 109.
104. E. Bülmann and H. Lund, *ibid.*, 1921, **16**, 321; 1923, **19**, 137.
105. F. Hovorka and W. C. Dearing, *J. Amer. Chem. Soc.*, 1935, **57**, 446.
106. S. Glasstone, *Introduction to Electrochemistry* (Van Nostrand, New York; Macmillan, London, 1942), pp. 291–5.
107. I. M.Kolthoff, *Rec. Trav. chim.*, 1923, **42**, 186.
108. V. Lester, *Agric. Science*, 1924, **14**, 634.
109. I. M. Kolthoff and W. Bosch, *Biochem. Z.*, 1927, **183**, 434.
110. E. B. R. Prideaux and F. T. Winfield, *Analyst*, 1930, **55**, 561.
111. A. Klit, *Z. phys. Chem.*, 1927, **131**, 61.
112. B. L. Clarke and L. A. Wooten, *J. Phys. Chem.*, 1929, **33**, 1468.
113. H. Seltz and D. S. McKinney, *Ind. Eng. Chem.*, 1928, **20**, 542.
114. B. L. Clarke and L. A. Wooten, *Ind. Eng. Chem. Analyt.*, 1930, **2**, 385.
115. H. S. Harned and D. D. Wright, *J. Amer. Chem. Soc.*, 1933, **55**, 4849..
116. J. L. R. Morgan, O. M. Lambert, and M. A. Cambell, *J. Amer. Chem. Soc.*, 1931, **53**, 454, 597, 2154; 1932, **54**, 910.
117. M. Cremer, *Z. Biol.*, 1906, **47**, 562.
118. F. Haber and Z. Klemenciewicz, *Z. phys. Chem.*, 1919, **67**, 385.
119. M. Dole, *The Glass Electrode* (John Wiley and Sons, New York, 1941).
120. D. A. MacInnes and M. Dole, *Ind. Eng. Chem. Analyt.*, 1929, **1**, 57.
121. *Idem, J. Amer. Chem. Soc.*, 1930, **52**, 29.
122. G. A. Perley, *Analyt. Chem.*, 1949, **21**, 391, 394, 559.
123. H. M. Partridge and J. A. C. Bowles, *Mikrochemie*, 1932, **11**, 326.
124. G. Schwarzenbach and W. Biedermann, *Helv. Chim. Acta*, 1948, **31**, 331; 456.
125. G. Schwarzenbach, *Analyt. Chim. Acta*, 1952, **7**, 141.
126. H. O. Day, J. S. Gill, E. V. Jones and W. L. Marshall, *Analyt. Chem.*, 1954, **26**, 611.
127. S. Bruckenstein and I. M. Kolthoff, *J. Amer. Chem. Soc.*, 1956, **78**, 10; 2974.
128. I. M. Kolthoff and S. Bruckenstein, *J. Amer. Chem. Soc.*, 1957, **79**, 1.
129. J. S. Fritz, *Acid-Base Titrations in Non-aqueous Solvents* (G. F. Smith Chemical Co., Columbus, Ohio, 1952).
130. *Idem, Analyt. Chem.*, 1954, **26**, 1701.
131. H. B. Van der Heijde, *Analyt. Chim. Acta*, 1957, **16**, 378; 392; 1957, **17**, 512.
132. A. J. Martin, *Analyt. Chem.*, 1958, **30**, 233.
133. R. Behrend, *Z. phys. Chem.*, 1893, **11**, 466.
134. W. D. Treadwell, *Helv. Chim. Acta*, 1919, **2**, 672.

135. A. LOTTERMOSER and W. PETERSEN, *Z. phys. Chem.*, 1928, **133**, 69.
136. E. LANGE and E. SCHWARTZ, *Z. Elektrochem.*, 1926, **32**, 240.
137. I. M. KOLTHOFF and L. VAN BERK, *Z. analyt. Chem.*, 1927, **70**, 369.
138. P. DUTOIT and G. VON WEISSE, *J. Chim. phys.*, 1911, **9**, 578.
139. E. LANGE and R. BERGER, *Z. Elektrochem.*, 1930, **36**, 980.
140. W. CLARK, *J. Chem. Soc.*, 1926, 768.
141. O. TOMIČEK, *Coll. Czech. Chem. Comm.*, 1930, **2**, 1.
142. J. J. LINGANE, *Analyt. Chem.*, 1954, **26**, 622.
143. R. L. KOWALKOWSKI, J. H. KENNEDY and P. S. FARRINGTON, *Analyt. Chem.*, 1954, **26**, 626.
144. E. MÜLLER, *Elektrometrische Massanalyse*, (T. Steinkopff, Dresden, 6. Auflage, 1942), p.111.
145. G. SPACU and P. SPACU, *Z. Analyt. Chem.*, 1934, **97**, 49.
146. I. M. KOLTHOFF and N. H. FURMAN, *Potentiometric Titrations* (John Wiley and Sons Inc., New York, 2nd. Edition, 1931), p. 159.
147. E. MÜLLER, *Z. anorg. Chem.*, 1924, **134**, 201.
148. A. MAZZUCHELLI and B. ROMANI, *Gazzetta*, 1927, **57**, 900.
149. M. H. BEDFORD, F. R. LAMB, and W. E. SPICER, *J. Amer. Chem. Soc.*, 1930, **52**, 583.
150. P. DUTOIT and G. VON WEISSE, *J. Chim. Phys.*, 1911, **9**, 630.
151. G. SCAGLIARINI and P. PRATESI, *Atti. Accad. naz. Lincei*, 1930, **11**, 193.
152. E. MÜLLER and W. LÖW, *Z. analyt. Chem.*, 1924, **64**, 297.
153. E. R. DAVIES and J. W. ARMSTRONG, *J. Inst. Petroleum*, 1944, **29**, 323.
154. J. H. KARCHMER, *Analyt. Chem.*, 1957, **29**, 425.
155. P. EKWALL and G. JUUP, *The Svedberg* (*Mem. Vol.*) 1944, 104.
156. E. SCHNELL, *Österr. Chem.-Ztg.*, 1953, **54**, 52.
157. M. M. RAINS and A. R. PUTNING, *Zavodskaya Lab.*, 1937, **6**, 888.
158. F. E. CRANE, *Analyt. Chem.*, 1956, **28**, 1794.
159. W. J. KIRSTEN, A. BERGGREN, and K. NILSSON, *Analyt. Chem.*, 1958, **30**, 237.
160. W. HILTER and W. GITTEL, *Z. analyt. Chem.*, 1934, **99**, 97.
161. J. DOLEZAL, V. HENCL, and V. SIMON, *Chem. Listy*, 1952, **46**, 267.
162. V. W. MELOCHE and L. KALBUS, *Analyt. Chem.*, 1956, **28**, 1047.
163. A. SCHULZE, *Z. analyt. Chem.*, 1957, **158**, 192.
164. D. G. BUSH, C. W. ZUEHLKE, and A. E. BALLARD, *Analyt. Chem.*, 1959, **31**, 1369.
165. W. D. TREADWELL and L. WEISS, *Helv. Chim. Acta*, 1919, **2**, 691.
166. E. MÜLLER and H. AARFLOT, *Rec. Trav. chim.*, 1924, **43**, 874.
167. C. MAYR and G. BURGER, *Monatsh.*, 1930, **56**, 113.
168. L. MARIQ, *Bull. Soc. chim. belges*, 1929, **38**, 426; 1930, **39**, 496; 1931, **40**, 361.
169. S. K. DUTTA and B. N. SHASH, *J. Indian. Chem. Soc.*, 1951, **28**, 383.
170. D. D. DEFORD and H. HORN, *Analyt. Chem.*, 1956, **28**, 797.
171. E. P. PRZYBYLOWICZ and L. B. ROGERS, *Analyt. Chem.*, 1956, **28**, 799.
172. *Idem, ibid.*, 1958, **30**, 1064.
173. I. M. KOLTHOFF, *Rec. Trav. chim.*, 1922, **41**, 172.
174. J. A. ATANASIU, *J. Chim. Phys.*, 1926, **23**, 501.
175. I. M. KOLTHOFF and E. J. A. VERZIJIL, *Rec. Trav. chim.*, 1924, **43**, 430.
176. J. J. LINGANE and A. M. HARTLEY, *Analyt. Chim. Acta*, 1954, **11**, 475.
177. D. G. DAVIS and H. T. MCLENDON, *Talanta*, 1959, **2**, 124.
178. E. MÜLLER and W. PRÉE, *Z. analyt. Chem.*, 1927, **72**, 195.
179. E. MÜLLER and K. GABLER, *Z. analyt. Chem.*, 1923, **62**, 29.
180. I. L. TEODOROVICH and M. A. ANDREEVA, *Zhur. analit. Khim.*, 1957, **12**, 100.
181. R. RIPON and E. POPPEV, *Z. analyt. Chem.*, 1948, **128**, 239.
182. A. RINGBOM, *Z. phys. Chem.*, 1935, **A173**, 198, 207.
183. I. SHUKOFF and T. G. RASKENSTEIN, *J. Gen. Chem.* (*U.S.S.R.*), 1934, **4**, 962.
184. R. W. FREEDMAN, *Analyt. Chem.*, 1959, **31**, 214.
185. M. LEBLANC and O. HARNAPP, *Z. phys. Chem.*, 1933, **A166**, 321.
186. J. VELÍŠEK, *Chem. Listy*, 1930, **24**, 443, 467.

187. C. A. Nierstrasz and H. J. C. Tendeloo, *Rec. Trav. chim.*, 1934, **53**, 792.
188. E. Denina and A. Caris, *Gazzetta*, 1938, **68**, 295.
189. R. Ripen-Tilici, *Z. analyt. Chem.*, 1938, **114**, 412.
190. E. M. Skobets and G. A. Kleibs, *J. Gen. Chem.* (*U.S.S.R.*), 1940, **10**, 1612.
191. I. M. Kolthoff and H. L. Sanders, *J. Amer. Chem. Soc.*, 193 7,**59**, 416.
192. H. J. C. Tendeloo and A. Krips, *Rec. Trav. chim.*, 1957, **76**, 703, 946.
193. R. B. Fischer and R. F. Babcock, *Analyt. Chem.*, 1958, **30**, 1732.
194. J. S. Parsons, *Analyt. Chem.*, 1958, **30**, 1262.
195. W. D. Treadwell, *Z. anorg. Chem.*, 1911, **71**, 223.
196. E. Müller and H. Lauterbach, *Z. anorg. Chem.*, 1922, **121**, 178.
197. W. Clark, *J. Chem. Soc.*, 1926, 768.
198. J. E. Ricci, *Analyt. Chem.*, 1953, **25**, 1650.
199. E. Müller and H. Lauterbach, *Z. analyt. Chem.*, 1922, **61**, 457.
200. T. Heczko, *Z. analyt. Chem.*, 1929, **78**, 325.
201. W. Bohnhaltzer, *Z. analyt. Chem.*, 1932, **87**, 401.
202. S. K. Chirkov and L. D. Naranovich, *J. Appl. Chem.* (*U.S.S.R.*), 1945, **18**, 699.
203. E. Müller and H. Lauterbach, *Z. analyt. Chem.*, 1923, **62**, 23.
204. E. Müller and A. Adam, *Z. Elektrochem.*, 1923, **29**, 49.
205. V. M. Tarayan, *Zavodskaya Lab.*, 1938, **7**, 176.
206. S. K. Chirkov, *Zavodskaya Lab.*, 1948, **14**, 783.
207. R. Flatt, *Helv. Chim. Acta*, 1934, **17**, 1494.
208. R. Přibil, Z. Koudela, and B. Matyska, *Coll. Czech. Chem. Comm.*, 1951, **16**, 80.
209. R. Patzak and G. Doppler, *Z. analyt. Chem.*, 1957, **156**, 248.
210. R. Belcher, D. Gibbons, and T. S. West, *Analyt. Chim. Acta*, 1955, **13**, 226.
211. L. Y. Polyak, *Zavodskaya Lab.*, 1955, **21**, 1300.
212. S. Siggia, D. W. Eichlen, and R. C. Rheinhart, *Analyt. Chem.*, 1955, **27**, 1745.
213. L. M. Budanova and O. Platanova, *Zavodskaya Lab.*, 1955, **21**, 1294.
214. L. Y. Polyak, *Zhur. analit. Khim.*, 1957, **12**, 224.
215. I. M. Kolthoff, *Analyt. Chem.*, 1954, **26**, 1685.
216. K. S. Knight and O. A. Osteryoung, *Analyt. Chim. Acta*, 1959, **20**, 481.
217. E. R. Nightingale, *Analyt. Chim. Acta.*, 1958, **19**, 587.
218. C. N. Reilley, *Sci. Apparatus and Methods* (E. H. Sargent Co.), 1957, **9**, 19.
219. F. Sadlek and C. N. Reilley, *Microchem. J.*, 1957, **1**, 185.
220. C. N. Reilley and R. W. Schmid, *Analyt. Chem.*, 1958, **30**, 947.
221. C. N. Reilley, R. W. Schmid and D. W. Lamson, *ibid.*, 953.
222. C. N. Reilley and A. Vavoulis, *Analyt. Chem.*, 1959, **31**, 243.
223. R. W. Schmid and C. N. Reilley, *J. Amer. Chem. Soc.*, 1956, **78**, 5513.
224. *Idem, Analyt. Chem.*, 1957, **29**, 264.
225. A. E. Martin and C. N. Reilley, *Analyt. Chem.*, 1959, **31**, 993.
226. N. Tanaka, I. T. Oiwa, and M. Kodama, *Analyt. Chem.*, 1956, **28**, 1555.
227. E. P. Przybylowicz and L. B. Rogers, *Analyt. Chem.*, 1958, **30**, 65.
228. I. M. Kolthoff and R. Belcher, *Volumetric Analysis*, (Interscience Publishers Inc., New York, 1957), Vol. III.
229. F. Crotogino, *Z. anorg. Chem.*, 1900, **24**, 224.
230. I. M. Kolthoff, *Rec. Trav. chim.*, 1921, **40**, 532.
231. F. L. Hahn and G. Weiler, *Z. analyt. Chem.*, 1926, **69**, 417.
232. F. L. Hahn, *Z. anorg. Chem.*, 1931, **195**, 75.
233. W. S. Hendrixson, *J. Amer. Chem. Soc.*, 1921, **43**, 858.
234. *Idem, ibid.*, 1309.
235. H. H. Willard and F. Fenwick, *J. Amer. Chem. Soc.*, 1923, **45**, 623.
236. J. A. Atanasiu, *Bull. Soc. Romana Stiinte*, 1928, **30**, 69.
237. W. S. Hendrixson, L. M. Verbeck, *Ind. Eng. Chem.*, 1922, **14**, 1152.
238. I. M. Kolthoff, *Pharm. Weekblad*, 1924, **61**, 841.
239. C. del Fresno, *Z. Elektrochem.*, 1925, **31**, 199.

240. G. L. KELLEY and R. T. BOHN, *J. Amer. Chem. Soc.*, 1919, **41**, 1776.
241. I. M. KOLTHOFF, *Rec. Trav. chim.*, 1922, **41**, 343.
242. E. J. A. VERZIJIL, *Dissertation*, Utrecht, 1923.
243. G. L. KELLEY, J. R. ADAMS and J. A. WILEY, *Ind. Eng. Chem.*, 1917, **9**, 780.
244. I. M. KOLTHOFF, *Chem. Weekblad*, 1919, **16**, 450.
245. E. MÜLLER and H. MOLLERING, *Z. anorg. Chem.*, 1924, **141**, 111.
246. D. T. EWING and E. F. ELDREDGE, *J. Amer. Chem. Soc.*, 1922, **44**, 1484.
247. R. G. GUSTAVSON and C. M. KNUDSON, *J. Amer. Chem. Soc.* 1922, **44**, 2756.
248. E. MÜLLER and A. FLATH, *Z. Elektrochem.*, 1923, **29**, 500.
249. W. D. TREADWELL, *Helv. Chim. Acta*, 1922, **5**, 732, 806.
250. E. MÜLLER, P. [illegible] G. UNGER, *Z. Elektrochem.*, 1927, **33**, 182.
251. E. MÜLLER and O. WAHLE, *Z. anorg. Chem.*, 1923, **129**, 33, 278; 1923, **130**, 63; 1923, **132**, 260.
252. R. MÜLLER and H. BRENNEIS, *Berg- u. Hüttenmann, Jahrb.-montan. Hochschule Leoben*, 1932, **80**, 101.
253. W. PUGH, *J. Chem. Soc.*, 1934, 1180.
254. O. STELLING, *Svensk kem. Tidskr.*, 1931, **43**, 130.
255. O. TOMIČEK, *Coll. Czech. Chem. Comm.*, 1939, **11**, 449.
256. J. J. LINGANE and R. KARPLUS, *Analyt. Chem.*, 1946, **18**, 191.
257. I. M. ISSA and S. AWAD, *Analyst*, 1953, **78**, 487.
258. M. BEZDEK and A. OKAC, *Chem. Listy*, 1951,**45**, 5.
259. C. O. HUBER and I. SHAIN, *Analyt. Chem.*, 1957, **29**, 1178.
260. I. M. ISSA and I. M. ELSHERIF, *Chemist-Analyst*, 1956, **45**, 78.
261. I. M. ISSA, R. M. ISSA and A. A. ABDUL, *Analyt. Chim. Acta*, 1954, **10**, 474.
262. C. S. FORBES and E. P. BARTLETT, *J. Amer. Chem. Soc.*, 1913, **35**, 1535.
263. K. SOMEYA, *Sci. Rep. Tohoku Univ.*, 1927, Series I, **16**, 303.
264. M. H. FLEYSHER, *J. Amer. Chem. Soc.*, 1924, **46**, 2725.
265. I. M. KOLTHOFF, *Rec. Trav. chim.*, 1920, **39**, 208.
266. *Idem, ibid.*, 1926, **45**,745.
267. F. K. FISHER,*J. Appl. Chem. (U.S.S.R.)*, 1936, **9**, 2269.
268. PROCTOR and GAMBLE CO., *Ind. Eng. Chem. Analyt.*, 1937, **9**, 514.
269. J. LÖBERING, *Z. analyt. Chem.*, 1935, **101**, 392.
270. S. D. GOKHALE, *J. Univ. Bombay*, 1952, Sec. A, **21**, 74.
271. H. F. LAUNER and Y. TOMIMATSU, *Analyt. Chem.*, 1953, **25**, 1767, 1769.
272. J. T. BYRNE, M. K. LARSEN and J. L. PFLUG, *Analyt. Chem.*, 1959, **31**, 942.
273. E. H. SWIFT and C. H. GREGORY, *J. Amer. Chem. Soc.*, 1930, **52**, 901.
274. H. H. WILLARD and P. YOUNG, *J. Amer. Chem. Soc.*, 1929, **51**, 139.
275. K. SOMEYA, *Z. anorg. Chem.*, 1939, **181**, 183.
276. N. H. FURMAN and J. H. WALLACE, *J. Amer. Chem. Soc.*, 1929, **51**, 1449.
277. H. H. WILLARD and P. YOUNG, *J. Amer. Chem. Soc.*, 1928, **50**, 1368.
278. E. H. SWIFT, *J. Amer. Chem. Soc.*, 1930, **52**, 894.
279. N. H. FURMAN, *J. Amer. Chem. Soc.*, 1928, **50**, 755.
280. H. H. WILLARD and P. YOUNG, *J. Amer. Chem. Soc.*, 1928, **50**, 1334.
281. *Idem, ibid.*, 1930, **52**, 553.
282. *Idem, ibid.*, 1928, **50**, 1379; 1929, **51**, 139.
283. D. T. EWING and M. WILSON, *J. Amer. Chem. Soc.*, 1931, **53**, 2105.
284. N. H. FURMAN and I. C. SCHOONOVER, *J. Amer. Chem. Soc.*, 1931, **53**, 2561.
285. N. H. FURMAN, *J. Amer. Chem. Soc.*, 1928, **50**, 1675.
286. H. H. WILLARD and P. YOUNG, *Ind. Eng. Chem.*, 1928, **20**, 972.
287. G. F. SMITH, *Cerate Oxidimetry* (G. F. Smith Chemical Co., Columbus, Ohio, 1942), Section 8.
288. D. G. STURGES, *Ind. Eng. Chem. Analyt.*, 1939, **11**, 267.
289. R. PŘIBIL and T. CHEBOVSKY, *Coll. Czech. Chem. Comm.*, 1947, **12**, 485.
290. R. PŘIBIL and V. MALICKY, *Coll. Czech. Chem. Comm.*, 1949, **14**, 413.
291. J. J. LINGANE, C. H. LANGFORD, and F. C. ANSON, *Analyt. Chim. Acta*, 1957, **16**, 271.

292. A. A. GRINBERG and A. I. DOBROBSOSKAYA, *Zhur. neorg. Khim.*, 1956, **1**, 2360.
293. I. M. KOLTHOFF, *Rec. Trav. chim.*, 1920, **39**, 212.
294. W. S. HENDRIXSON, *J. Amer. Chem. Soc.*, 1921, **43**, 861.
295. E. MÜLLER and D. JUNCK, *Z. Elektrochem.*, 1925, **31**, 200.
296. W. S. HENDRIXSON, *J. Amer. Chem. Soc.*, 1925, **47**, 1319.
297. F. FENWICK, *Dissertation*, Ann Arbor (1922), 76.
298. N. H. FURMAN and C. O. MILLER, *J. Amer. Chem. Soc.*, 1937, **59**, 152.
299. O. STELLING, *Svensk kem. Tidskr.*, 1933, **45**, 3.
300. I. C. SCHOONOVER and N. H. FURMAN, *J. Amer. Chem. Soc.*, 1923, **55**, 3123.
301. G. SPACU and C. DRAČULESCU, *Bull. Acad. Sci. Roumaine*, 1938, **20**, No. 1–3.
302. G. SPACU and P. SPACU, *Bull. Acad. Sci. Roumaine*, 1944, **26**, 295.
303. *Idem, Z. analyt. Chem.*, 1948, **126**, 233.
304. K. G. STONE, *Analyt. Chem.*, 1954, **26**, 396.
305. E. MISLOWITZER and W. SCHAEFER, *Biochem. Z.*, 1926, **168**, 203.
306. I. M. KOLTHOFF and J. J. VLEESCHHOUWER, *Rec. Trav. chim.*, 1926, **45**, 923.
307. E. ZINTL and H. WATTENBERG, *Ber.*, 1922, **55B**, 3366.
308. *Idem, ibid.*, 1923, **56**, 472.
309. E. ZINTL and K. BETZ, *Z. analyt. Chem.*, 1928, **74**, 341.
310. I. M. KOLTHOFF, *Rec. Trav. chim.*, 1920, **39**, 211.
311. E. ZINTL and G. RIENÄCKER, *Z. anorg. Chem.*, 1926, **153**, 276.
312. T. CALLAN and S. HORROBIN, *J. Soc. Chem. Ind.*, 1929, **47**, 334.
313. A. V. PAMFILOV and V. E. KISSELVA, *Z. analyt. Chem.*, 1927, **72**, 100.
314. T. NAKAZONO and S. INOKO, *Sci. Rep. Tohoku Univ.*, 1937, **26**, 303.
315. T. G. RAIKHINSTEIN, *Trans. Inst. Chem. Tech. Ivanovo* (*U.S.S.R.*), 1940, 17.
316. W. BIELENBERG and K. KÜHN, *Z. analyt. Chem.*, 1943, **126**, 88.
317. R. DOMANSKY, *Chem. Listy*, 1952, **46**, 480.
318. H. H. WILLARD and F. FENWICK, *J. Amer. Chem. Soc.*, 1923, **45**, 84.
319. I. M. KOLTHOFF and O. TOMIČEK, *Rec. Trav. chim.*, 1924, **43**, 447.
320. H. H. WILLARD and P. YOUNG, *Ind. Eng. Chem.*, 1928, **20**, 764, 769.
321. J. J. LINGANE and D. G. DAVIS, *Analyt. Chim. Acta*, 1956, **15**, 201.
322. K. SOMEYA, *Z. anorg. Chem.*, 1927, **168**, 56.
323. N. H. FURMAN, *J. Amer. Chem. Soc.*, 1928, **50**, 755.
324. E. MÜLLER and F. WEISBROD, *Z. anorg. Chem.*, 1928, **169**, 394.
325. C. R. N. STROUTS and C. A. MAC INNES, *Analyst*, 1948, **73**, 669.
326. O. NEIZOLDI, *Arch. Metallkunde*, 1949, **3**, 309.
327. R. WEINER, *Z. Elektrochem.*, 1948, **52**, 234.
328. L. C. W. BAKER and T. P. MCCUTCHEON, *Analyt. Chem.*, 1955, **27**, 1625.
329. F. KREJCI and L. KACETL, *Chem. and Ind.*, 1957, 598.
330. R. BELCHER, D. GIBBONS, and T. S. WEST, *Analyt. Chim. Acta*, 1955, **27**, 107.
331. T. PŘIBIL and J. DOLEŽAL, *Chem. Listy*, 1953, **47**, 1017.
332. R. W. SCHMID and C. N. REILLEY, *Analyt. Chem.*, 1956, **28**, 520.
333. W. D. TREADWELL, *Helv. Chim. Acta*, 1921, **4**, 396.
334. C. S. ROBINSON and O. B. WINTER, *Ind. Eng. Chem.*, 1918, **12**, 775.
335. E. ZINTL and P. ZAMIS, *Z. angew. Chem.*, 1927, **40**, 1288; 1928, **41**, 543.
336. R. LANG and J. ZWERINA, *Z. Elektrochem.*, 1928, **34**, 364.
337. *Idem, Z. analyt. Chem.*, 1932, **91**, 5.
338. I. M. KOLTHOFF and O. TOMIČEK, *Rec. Trav. chim.*, 1924, **43**, 798.
339. A. M. MCMILLAN and W. C. FERGUSON, *J. Soc. Chem. Ind.*, 1925, **44**, 141.
340. H. H. WILLARD and F. FENWICK, *J. Amer. Chem. Soc.*, 1923, **45**, 933.
341. E. ZINTL and G. RIENACKER, *Z. anorg. Chem.*, 1926, **155**, 84.
342. E. ZINTL and A. RAUCH, *Z. anorg. Chem.*, 1924, **39**, 397.
343. *Idem, Z. Elektrochem.*, 1925, **31**, 428.
344. E. ZINTL and G. RIENÄCKER, *Z. anorg. Chem.*, 1926, **153**, 278.
345. F. MÜLLER, *Z. analyt. Chem.*, 1926, **69**, 168.
346. W. D. TREADWELL and M. ZÜRCHER, *Helv. Chim. Acta*, 1927, **10**, 281.
347. W. R. CROWELL and H. D. KIRSCHMAN, *J. Amer. Chem. Soc.*, 1929, **51**, 1695.

348. J. J. LINGANE and J. H. KENNEDY, *Analyt. Chim. Acta*, 1956, **15**, 465.
349. S. C. WOO and D. M. YOST, *J. Amer. Chem. Soc.*, 1931, **53**, 884.
350. W. S. HENDRIXSON, *J. Amer. Chem. Soc.*, 1923, **45**, 2013.
351. I. M. KOLTHOFF and O. TOMIČEK, *Rec. Trav. chim.*, 1924, **43**, 788.
352 T. CALLAN and S. HORROBIN, *J. Soc. Chem. Ind.*, 1928, **47**, 333.
353. V. A. MATULA, *Chem. Obzor.*, 1931, **6**, 124.
354. E. ZINTL and F. SCHLOFFER, *Z. angew. Chem.*, 1928, **41**, 956.
355. E. ZINTL and G. RIENÄCKER, *Z. anorg. Chem.*, 1927, **161**, 385.
356. E. ZINTL, G. RIENÄCKER and F. SCHLOFFER, *ibid.*, 1928, **168**, 97.
357. J. J. LINGANE and C. AUERBACH, *Analyt. Chem.*, 1951, **23**, 986.
358. H. BRINTZINGER and F. RODIS, *Z. anorg. Chem.*, 1927, **166**, 53.
359. E. ZINTL and P. ZAMIS, *Z. angew. Chem.*, 1927, **40**, 1286.
360. H. BRINTZINGER and F. OSCHATZ, *Z. anorg. Chem.*, 1927, **165**, 221.
361. H. BRINTZINGER and B. RAST, *Z. analyt. Chem.*, 1939, **115**, 241; **117**, 1; 1940, **120**, 165.
362. W. R. CROWELL and H. L. BRUMBACH, *J. Amer. Chem. Soc.*, 1935, **57**, 2607.
363. R. FLATT and F. SOMMER, *Helv. Chim. Acta*, 1942, **25**, 684, 1518; 1944, **27**, 1522.
364. J. J. LINGANE and L. NIEDRACH, *J. Amer. Chem. Soc.*, 1948, **70**, 1997.
365. R. S. BOTTEI and N. H. FURMAN, *Analyt. Chem.*, 1955, **27**, 1182; 1957, **29**, 121.
366. V. W. MELOCHE and R. L. MARTIN, *Analyt. Chem.*, 1956, **28**, 1671.
367. E. MÜLLER and J. GÖRNE, *Z. analyt. Chem.*, 1928, **73**, 385.
368. E. MÜLLER and W. STEIN, *Z. Elektrochem.*, 1930, **36**, 376.
369. E. MÜLLER and R. BENNEWITZ, *Z. anorg. Chem.*, 1929, **179**, 113.
370. E. MÜLLER and J. GÖRNE, *Z. analyt. Chem.*, 1928, **73**, 383.
371. W. TRZEBIATOWSKI, *Z. analyt. Chem.*, 1930, **82**, 45.
372. E. MÜLLER and G. HASSE, *Z. analyt. Chem.* 1933, **91**, 241.
373. I. MARTYNCHENKO and A. SHIMKO, *Zavodskaya Lab*, 1936, **5**, 1297.
374. H. HÖLEMAN, *Z. anorg. Chem.*, 1934, **217**, 33.
375. D. V. VASILEV, *J. Appl. Chem.* (*U.S.S.R.*), 1941, **14**, 689.
376. B. NEUMANN and G. MEYER, *Z. analyt. Chem.*, 1949, **129**, 229.
377. A. J. BARD and J. J. LINGANE, *Analyt. Chim. Acta*, 1959, **20**, 463.
378. W. D. TREADWELL and R. NIERIKER, *Helv. Chim. Acta*, 1941, **24**, 1067, 1098.
379. G. F. SMITH and C. A. GETZ, *Ind. Eng. Chem. Analyt.*, 1938, **10**, 191.
380. E. WADSWORTH, F. R. DUKE, and C. A. GOETZ, *Analyt. Chem.*, 1957, **29**, 1824.
381. H. H. WILLARD and P. YOUNG, *J. Amer. Chem. Soc.*, 1928, **50**, 1322.
382. R. B. HAHN and M. T. KELLEY, *Analyt. Chim. Acta*, 1954, **10**, 178.
383. G. R. WATERBURY and C. F. METZ, *Analyt. Chem.*, 1959, **31**, 1144.
384. A. J. FENTON and N. H. FURMAN, *Analyt. Chem.*, 1957, **29**, 221.
385. N. H. FURMAN, *Analyt. Chem.*, 1953, **25**, 482.
386. V. P. RAO and G. G. RAO, *Talanta*, 1958, **1**, 335.
387. D. MONNIER and P. ZWAHLEN, *Helv. Chim. Acta*, 1956, **39**, 1859.
388. I. M. KOLTHOFF and E. R. NIGHTINGALE, *Analyt. Chim. Acta*, 1959, **19**, 593.
389. I. M. KOLTHOFF and A. LAUR, *Z. analyt. Chem.*, 1928, **73**, 177.
390. F. L. HAHN, *Z. angew. Chem.*, 1927, **40**, 351.
391. C. DEL FRESNO and L. VALDES, *Z. anorg. Chem.*, 1929, **183**, 251.
392. O. TOMIČEK, *Rec. Trav. chim.*, 1925, **44**, 410.
393. O. TOMIČEK and F. FRESBERGER, *J. Amer. Chem. Soc.*, 1935, **57**, 1209.
394. O. TOMIČEK and J. KALNY, *J. Amer. Chem. Soc.*, 1935, **57**, 801.
395. A. RIUS and C. A. DIAZ-FLORES, *Anales real Soc. españ. Fis. Quim*, 1950, **46B**, 289.
396. G. T. HOIJTINK, E. DE BOER, P. H. VAN DER MEIJ, and W. P. WEIJLAND, *Rec., Trav. chim.*, 1956, **75**, 487.
397. G. L. KELLEY and J. B. CONANT, *Ind. Eng. Chem.*, 1916, **8**, 719; 1917, **9**, 780; 1921, **13**, 1053.
398. K. L. CHENG, *Analyt. Chem.*, 1955, **27**, 1165.
399. R. PŘIBIL, J. DOLEŽAL and V. SIMON, *Coll. Czech. Chem. Comm.*, 1951, **16**, 573.

400. C. D. Wagner, R. H. Smith and E. D. Peters, *Analyt. Chem.*, 1947, **19**, 982.
401. W. E. Shaefer and W. W. Becker, *Analyt. Chem.*, 1953, **25**, 1226.
402. J. J. Lamond, *Analyt. Chim. Acta*, 1953, **8**, 217.
403. R. H. Pierson and E. Gantz, *Analyt. Chem.*, 1954, **26**, 1809.
404. I. M. Kolthoff, *Pharm. Weekblad*, 1922, **59**, 66.
405. J. H. Kennedy and J. J. Lingane, *Analyt. Chim. Acta*, 1958, **18**, 240.
406. P. Arthur and J. F. Donahue, *Analyt. Chem.*, 1952, **24**, 1612.
407. E. Müller and W. Stein, *Z. Elektrochem.*, 1930, **36**, 220.
408. W. G. Berl, *Physical Methods in Chemical Analysis* (Academic Press, Inc., New York, 1951), Vol. II, p. 149.
409. T. F. Buehrer and O. E. Schupp, *Ind. Eng. Chem.*, 1926, **18**, 121.
410. J. J. Lingane and R. L. Pecsok, *Analyt. Chem.*, 1948, **20**, 425.
411. *Idem, ibid.*, 1949, **21**, 622.
412. A. J. Bard and J. J. Lingane, *Analyt. Chim. Acta*, 1959, **20**, 581.
413. J. J. Lingane, *Analyt. Chim. Acta*, 1959, **21**, 227.
414. K. Maass, *Z. analyt. Chem.*, 1934, **97**, 241.
415. P. C. Banerjee, *J. Indian Chem. Soc.*, 1938, **15**, 475.
416. M. Matrka and Z. Sagner, *Coll. Czech. Chem. Comm.*, 1957, **22**, 1131.
417. P. C. Banerjee, *J. Indian Chem. Soc.*, 1942, **19**, 30, 35.
418. G. L. Kelley, M. G. Spencer, C. B. Illingworth and T. Gary, *Ind. Eng. Chem.*, 1918, **10**, 19.
419. R. Belcher and T. S. West, *Analyt. Chim. Acta*, 1952, **7**, 470.
420. R. Belcher, D. Gibbons and T. S. West, *Analyt. Chem.*, 1954, **26**, 1025.
421. L. Erdey and G. Rody, *Talanta*, 1958, **1**, 159.
422. W. M. Thornton, jr., and E. Chapman, *J. Amer. Chem. Soc.*, 1921, **43**, 91.

Chapter IV

Conductometric Titrations

DONALD G. DAVIS

1. Introduction

The equivalence point of a great many titrations may be found by measuring the conductance of the solution titrated as a function of the volume of standard titrant added; in fact, resistance is the variable most often measured. The resistance, R, in ohms, of a column of an electrolytic solution contained between two identical electrodes of area A cm², is given by

$$R = \varrho\,(L/A), \qquad (1)$$

where L is the length of the column of electrolyte, and ϱ is a proportionality constant called the specific resistance.

The reciprocal of the resistance is the conductance in reciprocal ohms, designated mhos

$$\frac{1}{R} = \frac{A}{L}\varkappa, \qquad (2)$$

the proportionality constant now being $\varkappa$, the specific conductance.

If the electrolytic solution under consideration behaves ideally, the specific conductance is a linear function of the electrolyte concentration, c, and this linear dependence on concentration can be used for direct concentration measurements, provided that the concentration of only one electrolyte is measured. In conductometric titrations, the situation is more complex, because at least two electrolytes are involved; several more relationships and definitions must therefore be considered.

The molar conductance is defined as the conductance of a cell with electrodes 1 cm apart and of such dimensions that one mole of electrolyte is contained in the solution between them; that is, if the concentration is expressed in moles per 1000 ml,

$$A = \frac{1000}{c}, \qquad (3)$$

as $L = 1$ cm, equation (2) then becomes

$$\Lambda_m = \frac{1000}{c} \varkappa, \tag{4}$$

Λ_m signifying the molar conductance. If c is expressed in units of equivalents per 1000 ml, the equivalent conductance, Λ, is given by:

$$\Lambda = \frac{1000}{c} \varkappa. \tag{5}$$

The conductance can now be expressed in terms of Λ and c.

$$\frac{1}{R} = \frac{\Lambda c A}{1000\ L} \tag{6}$$

In a great many of the cells used in practice, the electrolyte surrounds the electrodes, and the electrical field is thus not uniform. As the ratio L/A cannot be obtained by simple geometry, this ratio, designated θ, must be measured for each cell used, if absolute results are desired; in conductometric titrations, relative values are, however, usually sufficient and θ need not be known.

$$\theta = \varkappa R = \frac{\Lambda c R}{1000} \tag{7}$$

In order to calibrate a cell, a standard solution of exactly known specific conductance must be available. Parker and Parker[1], and Jones and Bradshaw[2] have made very careful determinations of the specific conductance of potassium chloride solutions, and some of the results of Jones and Bradshaw are recorded in Table IV, 1. It is considered advisable to determine

TABLE IV, 1

SPECIFIC CONDUCTANCE OF SOLUTIONS OF POTASSIUM CHLORIDE

g KCl per 1000 g of solution in vacuum	$\varkappa$ *ohm^{-1} cm^{-1}* 18°	25°
71.1352	0.09784	0.11134
7.41913	0.011167	0.012856
0.745263	0.0012205	0.0014088

the cell constant with a standard solution whose specific conductance is of a similar value to that of the solution to be investigated.

As previously indicated, the specific conductance for an ideal electrolyte should be proportional to the number of ions present per unit volume;

References p. 211

likewise, Λ and Λ_m should be absolutely independent of concentration. In practice, however, Λ and Λ_m generally decrease as the concentration increases. This phenomenon is indicated in Fig IV, 1, which shows a plot of the ratio Λ/Λ_0 against concentration. Here, Λ is the equivalent conductance at any finite concentration, and Λ_0 is the equivalent conductance at infinite dilution.

Curves similar to curve 2 are obtained with strong electrolytes such as sodium chloride and nitric acid, which are generally considered to be completely dissociated. The decrease in Λ with increasing concentration is due

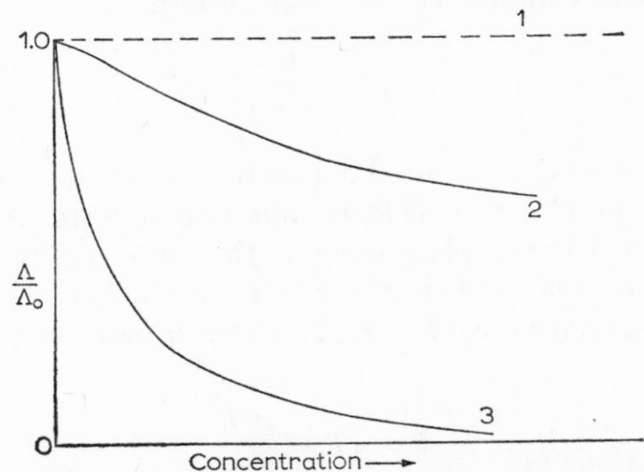

Fig. IV, 1. Variation of equivalent conductance with concentration: 1. Ideal electrolytes. 2. Strong electrolytes. 3. Weak electrolytes.

to the limiting of the freedom of movement of the ions by the electrostatic attraction of neighbouring ions, so that the solution behaves as if it contained less than the actual number of ions present. Curve 3 is for weak electrolytes, which only approach complete ionisation at infinite dilution.

Each ion in solution makes an individual contribution to the total conductance,

$$\frac{1}{R} = \frac{1}{1000\,\theta} \Sigma C_i l_i, \tag{8}$$

where l_i is the equivalent ionic conductance of the ith ion. If only a single salt is involved,

$$\Sigma C_i l_i = C_+ l_+ + C_- l_-, \tag{9}$$

the units of the C's being in equivalents per litre.

The values of equivalent ionic conductance can be obtained through the transference (transport) number, N, because

$$N_i = \frac{l_i}{\Lambda_{salt}} \tag{10}$$

Transference numbers can be determined by the classical Hittorf method, or by measurement of the e.m.f. of the appropriate "concentration cell with transference".

Table IV, 2 lists a number of equivalent ionic conductances at infinite dilution at 25°. These may be used to predict the form of any particular titration curve, as will be seen in subsequent sections of this chapter.

TABLE IV, 2

EQUIVALENT IONIC CONDUCTANCES[3,4] AT 25°

Cations	l_0	*Anions*	l_0
H^+	350	OH^-	198
Tl^+	75	Br^-	78
K^+	74	I^-	77
NH_4^+	73	Cl^-	76
Ag^+	62	NO_3^-	71
Na^+	50	ClO_4^-	67
Li^+	39	ClO_3^-	64
$\frac{1}{2}Pb^{2+}$	73	BrO_3^-	56
$\frac{1}{2}Ba^{2+}$	64	F^-	55
$\frac{1}{2}Ca^{2+}$	60	$HCOO^-$	55
$\frac{1}{2}Sr^{2+}$	59	IO_4^-	55
$\frac{1}{2}Fe^{2+}$	54	HCO_3^-	44
$\frac{1}{2}Cu^{2+}$	54	CH_3COO^-	41
$\frac{1}{2}Mg^{2+}$	53	$C_6H_5COO^-$	32
$\frac{1}{2}Zn^{2+}$	53	$B(C_6H_5)_4^-$	21
$\frac{1}{2}UO_2^{2+}$	51	$\frac{1}{2}CrO_4^{2-}$	82
$\frac{1}{3}Co(NH_3)_6^{3+}$	102	$\frac{1}{2}SO_4^{2-}$	80
$\frac{1}{3}Fe^{3+}$	68	$\frac{1}{2}CO_3^{2-}$	69
		$\frac{1}{2}C_2O_4^{2-}$	24
		$\frac{1}{3}Fe(CN)_6^{3-}$	101
		$\frac{1}{3}PO_4^{3-}$	80
		$\frac{1}{4}Fe(CN)_6^{4-}$	111

It should be borne in mind that the resistance, and thus the conductance, of an electrolytic solution is profoundly effected by changes in temperature: the resistance decreases by about 1% to 2% for each degree increase in temperature; for this reason, it is desirable to carry out a conductometric titration at an approximately constant temperature. If absolute measurements are to be made, a constant-temperature bath is of course necessary.

Although the equivalent ionic conductances given in Table IV, 2 were measured at 25°, approximate correction at any temperature may be obtained by use of the fact that for most ions $1/l_0\ (dl_0/dT)$ is close to 0.02.

2. Apparatus and Techniques

(a) Measuring Circuits

In order to prevent concentration changes due to reaction at the electrodes, the resistance of solutions is generally measured with alternating current; frequencies from 60 to 10,000 c/sec have been used for ordinary conductance measurements, although work at high frequencies finds many applications (see Chapter V). Most of the circuits used are of the Wheatstone-bridge type illustrated in Fig. IV, 2: R_1, R_2, and R_3 are precision variable resistors; pure sine wave alternating current is applied to points A and C and some type of a.c. null detector (N.D.) is connected across points B and D; a galvanometer, or an ordinary telephone receiver, is the classical null detector, but more convenient devices, such as the cathode ray tube, are now used.

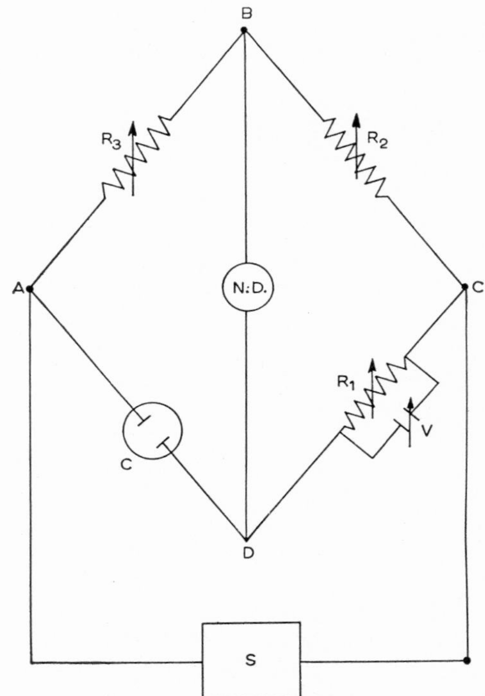

Fig. IV, 2. Wheatstone bridge.

When the bridge is balanced, points B and D are at the same potential so that:

$$i_{ABC}\, R_3 = i_{ADC}\, R_{cell} \tag{11}$$

$$i_{ABC}\, R_2 = i_{ADC}\, R_1 \tag{12}$$

and

$$R_{cell} = R_3\, R_1/R_2 \tag{13}$$

Actually, at the balance point the impedances, I, rather than the resistances, are balanced.

$$I_{cell} = I_3\, I_1/I_2 \tag{14}$$

The impedance is related to the resistance and the reactance, X, by

$$I = \sqrt{R^2 + X^2} \tag{15}$$

The reactance depends on the inductance and the capacitance, and for a series-type circuit is given by

$$X = \left(2\,\pi fL - \frac{1}{2\pi\, fC}\right), \tag{16}$$

where f is the frequency in c/sec, L is the inductance in henries and C is the capacitance in farads. It can be seen that equation (13) is only valid as long as

$$\frac{X_{cell}}{R_{cell}} = \frac{X_1}{R_1} \text{ and } \frac{X_3}{R_3} = \frac{X_2}{R_2}. \tag{17}$$

In practice, the apparatus is designed to make the inductance negligibly small, but the capacitance of the cell cannot be entirely eliminated, and this causes the null detector to be somewhat insensitive; in accurate work, the cell capacity is balanced by use of the variable condenser connected across R_1. The resistances R_2 and R_3 commonly consist of a slide-wire and R_1 is a resistance box; if the slide-wire is divided into a thousand arbitrary units and the point of contact is represented by x,

$$\frac{R_3}{R_2} = \frac{x}{1000 - x}. \tag{18}$$

Balance is obtained by varying both the resistance box and the slide-wire, and the greatest accuracy is obtained when $x = 1000 - x$. Much of the older work on conductometric titrations shows titration curves for which the value of $(1000 - x)/x$ is plotted against volume.

Although the basic instrument for measuring conductivity is that shown in Fig. IV, 2, the literature contains a large number of variations, which

References p. 211

usually take the form of new a.c. sources or null detectors. Induction coils were first used as a.c. sources, but high-frequency generators and, in particular, oscillating thermionic valves[5, 6, 7, 8], have found extensive application, and are highly recommended[9]. A constant-voltage transformer connected directly to the a.c. supply makes a very simple source[10]; it can only be used for conductometric titrations, and not for the measurement of absolute conductivities.

As audio methods are usually tedious, null detectors have received considerable attention: an a.c. galvanometer[11], or a d.c. galvanometer plus a rectifier[12], is often recommended; valve voltmeters[13] and simple a.c. milliammeters[10] have also been suggested; even a photoelectric colorimeter[14] has been used as a null-point detector by connecting an electric lamp in series with the solution and, by the use of a photocell, monitoring the changes in its emission caused by changes in the resistance of the solution.

Several types of automatic recording or titrating devices have been suggested: for instance, Juliard and van Cakenberghe[15] designed an apparatus consisting of a titration cell in a balanced Wheatstone-bridge circuit with a recording millivoltmeter as a null detector; as the titrant is added from a screw-driven syringe, the displacement is recorded by the meter. Goodwin[16] has described a continuously recording conductivity meter that could, presumably, be used for titrations.

Stock[15] has published an excellent and critical review of the apparatus sued for conductometric titrations, especially micro titrations.

(b) Commercial Apparatus

Fairly simple commercially available equipment is accurate enough for most conductometric titrations; its main advantages are that it is convenient and can be set up and used without any knowledge of electricity or electronics. The normal type of commercial apparatus is exemplified by the Industrial Instruments, Inc. (Cedar Grove, N.J., U.S.A.) Type RC instruments (Fig. IV, 3). These wide-range instruments are essentially null-balance Wheatstone bridges with low-voltage valve amplifiers and electron ray "magic eye" tubes as null-point indicators; to make a measurement, the operator simply brings the bridge into balance. On most models, either 60 c/sec or 1000 c/sec may be selected, according to whether the solution to be measured has a high or a low resistance. Ranges of 0.2–2,500,000 ohms and 0.4–5,000,000 micro-ohms are available, and ohms or mhos may be read from the scale. Instruments of this type are also available from Wissenschaftlich-Technische-Werkstatten (Weilheim 10 Berbayem, Germany) and Deutsche Metrohm (Essen/Ruhr, Germany).

The less conventional type of commercial apparatus is the direct-reading

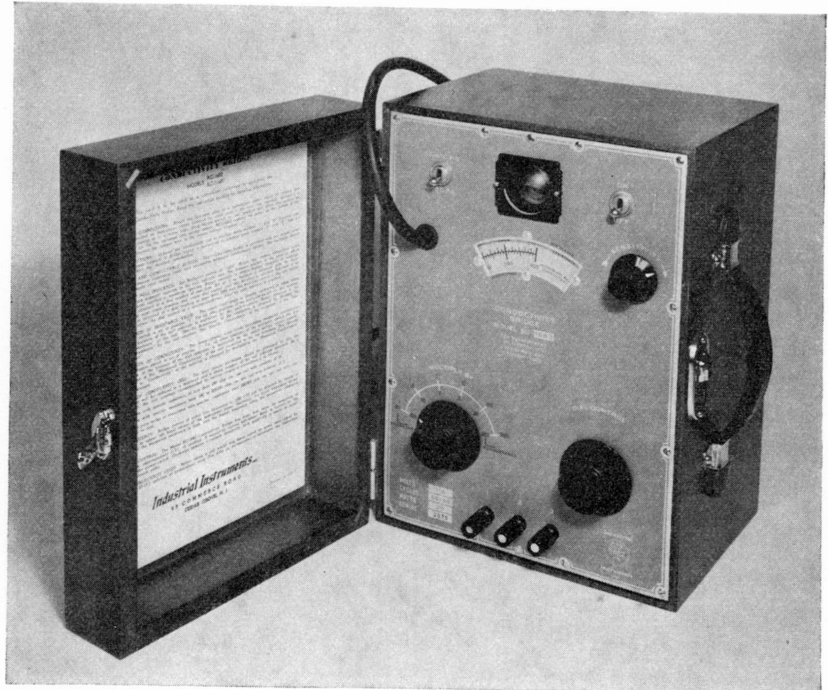

Fig. IV, 3. Industrial Instruments conductivity bridge.
(Courtesy Industrial Instruments Inc.)

instrument, such as the CDM2 meter made by Radiometer (Copenhagen NV, Denmark): the principle of this meter is very similar to that of an ohmmeter, except that the scale is calibrated in reciprocal ohms and the test voltage is alternating, with a frequency of either 70 or 300 c/sec; the meter has a mirrored scale and a knife-edge pointer for greater accuracy. Another maker of this type of instrument is J. Tacussel (Lyon/3, France), whose product, which is illustrated in Fig. IV, 4, is particularly stable to changes in mains voltage.

The meter type of apparatus is much more convenient for titration purposes, but it is probably not as accurate as the bridge type. High accuracy is not necessary for most titrations, and even precision can be sacrificed to some extent, because, when conductometric titration curves are plotted, a graphical averaging process is actually effected. When high accuracy ($\pm$ 0.001%) is needed, or for single conductivity measurements, an instrument such as the Jones and Joseph[7] bridge made by the Leeds and Northrup Co. (Philadelphia, Pa., U.S.A.) should be used.

References p. 211

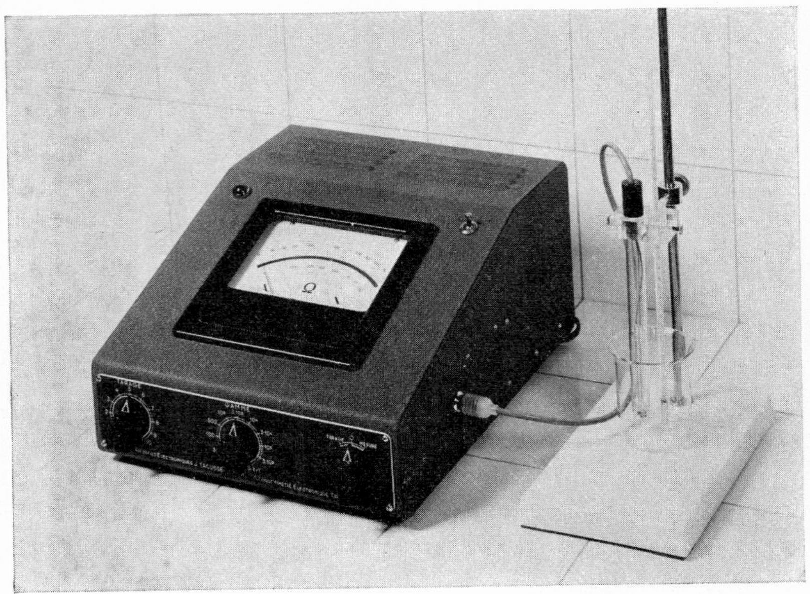

Fig. IV, 4. Direct-reading conductivity bridge
(Courtesy of the Société Lyonnaise d'Electronique Appliquée.)

(c) Electrodes and Cells

A number of titration cells have been recommended, of which those of Britton[17] and of Kano[18] are the most practical. In addition, dipping electrodes, such as that illustrated in Fig. IV, 5, are available in various sizes from the makers of the instruments discussed in the previous section. The simplest titration cell consists of a dipping electrode, a beaker of the appropriate size and a mechanical stirring device. The electrodes should be rigidly mounted, parallel to each other, and such a distance apart that reasonable values of resistance are measured; often, several sets of dipping electrodes are kept at hand for use in different titrations. For precipitation titrations, the electrodes are mounted vertically to prevent excessive amounts of precipitate from collecting on the electrode surfaces.

Because a large effective surface area minimises polarisation, most conductivity electrodes are plated with a layer of platinum black; the platinisation of electrodes also tends to decrease the cell capacitance and allows the establishment of a sharp balance point. On the other hand, platinised electrodes have a great tendency to adsorb substances from solution, and difficulties are encountered with dilute solutions or solutions containing surface-active agents; bright platinum or lightly platinised electrodes are used in such cases.

A few unconventional electrode systems have also been recommended: for example, Khlopin[19] carried out conductometric titrations with the cell,

$$Zn/ZnSO_4//\text{test solution}/\text{graphite},$$

but this type of cell has found only limited application.

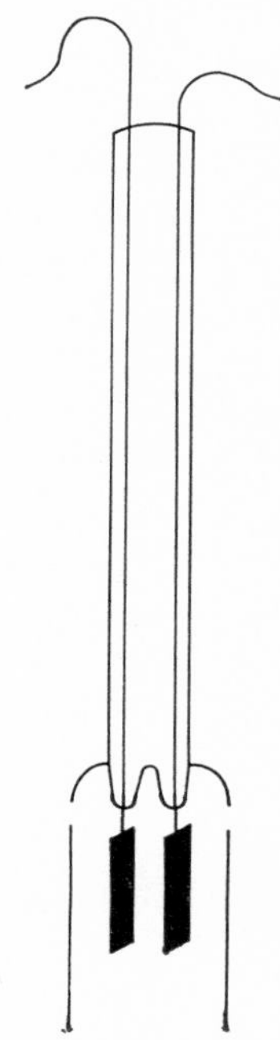

Fig. IV, 5. Dipping conductivity cell.

(d) Techniques

Accurate control of temperature is important in conductance measurements (page 177) but, for conductometric titrations, constancy of temperature is all that is required, and a thermostatted water bath is unnecessary; the cell is usually placed in a large pan of water at room temperature to eliminate fluctuations due to heats of reaction or outside causes. An automatic temperature control devised by Müller and Vogel[20] allows the measurement of the concentration of a single salt to 0.1%. Very effective temperature compensation for titrations can be achieved by using two cells, one in each of two branches of the Wheatstone bridge[13].

References p. 211

Because the volume increases during a conductometric titration, the resulting titration graphs do not consist of straight lines, and this may cause some difficulty in locating the end point; hence, the observed conductance values are often corrected for volume change before they are plotted. The correction factor is $(V + v)/V$, where V is the initial volume and v is the volume of titrant added; this assumes that conductivity is a linear function of dilution, which is not strictly true for real solutions, but the correction will be accurate enough if v is small compared to V. For this reason, conductometric titrations are best carried out with a titrant at least ten times more concentrated than the solution being titrated.

(e) Special Methods

Although conductance is generally measured by means of a.c. apparatus, it has been shown by a number of workers that d.c. can be used, provided that the cell has two unpolarisable electrodes, such as the Ag/AgCl electrode. Eastman[21] has shown that the direct-current method is capable of yielding results that agree to 0.01% with those obtained with the usual a.c. methods. Taylor and Furman[22] have recommended a d.c. method for conductometric titrations: their cell contains two platinum electrodes through which a known constant current is passed; two tungsten probe electrodes are placed in the field between the platinum electrodes, and show a difference in potential that is proportional to iR, the product of the constant current and the resistance of the solution; as the potential can be measured with a pH meter or some other potential measuring device, the apparatus is quite simple. Although this technique is subject to errors due to concentration changes in the solution resulting from the electrode reactions, the errors can be reduced to negligible size by keeping the current small (microamps) and only allowing it to pass when the potential difference is being measured.

Differential conductometric titrations have been investigated by Duval and Duval[23], who used two cells connected in two branches of a Wheatstone bridge; both cells contained identical sample solutions and both were titrated with the same reagent, but one cell always had 0.05 to 0.1 ml more titrant in it than the other. The titration curves for each cell were plotted as one graph and their point of intersection gave the end point; this method has been applied to the micro titration of zinc with ferrocyanide, calcium with oxalate, magnesium with phosphate and permanganate with iron[24]. Delahay[25] has devised a differential indicator-current method in which a constant alternating current is fed to the cell from a regulated power supply. The alternating potential across the cell is rectified and applied to a condenser, and the resulting current is measured with a microammeter. The method is rapid and reproducible, and does not require temperature control or volume correction, and titrations can be carried out in boiling solutions; it can, however, only be used if the titration curve is characterised by a maximum or a minimum.

3. Acid-Base Titrations

(a) Conductometric Titration of Strong Acids and Strong Bases

A typical strong acid/strong base titration may be exemplified by

$$H^+ + Cl^- + Na^+ + OH^- = Na^+ + Cl^- + H_2O, \tag{19}$$

the net result being the replacement of hydrogen ions by sodium ions. As Table IV, 2 shows, the equivalent ionic conductance of the hydrogen ion is

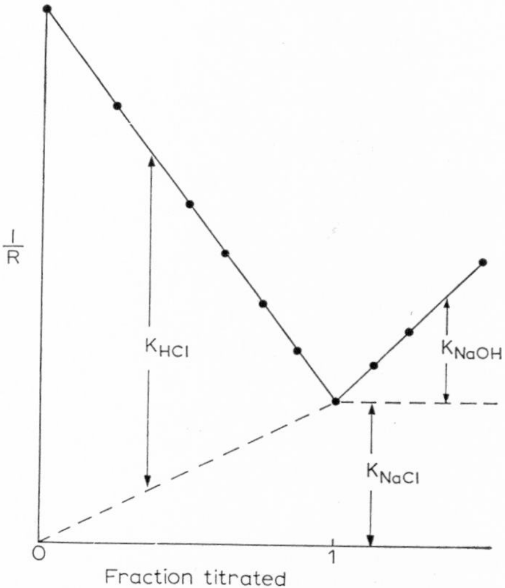

Fig. IV, 6. Conductometric titration curve of a strong acid with a strong base. (A/L being equal to unity)

much greater than that of the sodium ion, so that the conductance of the solution at the beginning of the titration is large but decreases up to the equivalence point; this is shown in Fig. IV, 6. At the equivalence point, the conductance is equal to that of a pure solution of sodium chloride and is at a minimum. Beyond the equivalence point, the conductance increases in direct proportion to the excess of sodium hydroxide added.

The slope of the first branch of the titration curve, and thus the sharpness with which the end point may be established, depends on the difference

References p. 211

between the equivalent ionic conductance of the hydrogen ion and the cation of the base used, in this case the sodium ion. As shown in Table IV, 2, the equivalent ionic conductance of the hydrogen ion is much greater than that of any other cation; because of this, strong acids can almost always be successfully titrated to a conductometric end point.

One especially important advantage of conductometric titrations is that the relative change in conductance is almost independent of the concentration of the strong acid titrated; thus, dilute solutions can be titrated with almost the same accuracy as concentrated solutions; the titration curve (Fig. IV, 6) is displaced along the $1/R$ axis, but the sharpness is not greatly affected. In this respect, conductometric acid-base titrations differ from potentiometric titrations with a glass electrode.

Treadwell[26] has pointed out that, because of the greater mobility of the hydrogen ion as compared with that of the hydroxyl ion, the point of minimum conductivity does not lie at exactly the same point on the titration curve as true neutrality; the point of minimum conductivity has been calculated to be at pH 7.1, but the difference between this and the theoretical pH 7.0 is negligible.

Kolthoff[27] and Dutoit[28, 29] have shown that $10^{-4}M$ solutions of hydrochloric acid can be accurately titrated if carbon dioxide is excluded. The specific conductance at the equivalence point would be about 1.3×10^{-5} mho, while ordinary distilled water containing carbon dioxide would have a specific conductance of about 0.2×10^{-5} mho. In practice, the conductivity of the solvent would cause no error in a conductometric titration if it remained constant but, because carbon dioxide behaves as a weak acid, an excess of standard base would be consumed and the titration curve would not be sharp. For work at ordinary concentrations, carbon-dioxide-free water is not necessary, provided that the strong acid is titrated with a strong base, rather than the reverse[30].

Although the titration of strong acids with strong bases was among the first conductometric titrations attempted[31], the conductometric method is not much used for such determinations, because a large number of indicator methods are available; even coloured solutions are now mostly titrated potentiometrically with a glass electrode.

One specially interesting application is the determination of sulphuric and nitric acids in nitrating acids: Müller and Kogert[32] first titrated the total hydrogen ion with sodium hydroxide and then titrated the neutralised solution with lead nitrate after adding 30–40% of ethanol; lead sulphate was thus precipitated efficiently, and the titration gave the amount of sulphuric acid present; the nitric acid was found by difference. Later, Clark[33] recommended that, after the total acidity had been determined, the nitric acid should be removed from another sample and the remaining sulphuric acid titrated with a standard solution of barium acetate in hot aqueous propanol; the nitric acid was destroyed by treatment with formaldehyde in the presence of an excess of methanol, the resulting low-boiling methyl formate and the excess of methanol being removed by distillation.

Conductometric titrations seem particularly advantageous for the titration of strong acids in the presence of hydrolysable ions: Murgulescu and Latiu[34] have titrated sulphuric acid in the presence of aluminum sulphate and Pepkowitz *et al.*[35] have determined free hydrogen ions in solutions containing iron, aluminum, chromium, nickel and magnesium with the help of fluoride as a complexing agent; the method of these workers is outlined below.

Determination of Free Acid in the Presence of Hydrolysable Cations

Apparatus:

Any normal bridge circuit. A small titration cell of about 40 ml capacity, a small clip-type of electrode assembly and a magnetic stirrer.

(Procedure:

Add an appropriate sample, 100 microlitres or less, to 20 ml of distilled water in the titration cell, followed by the exact amount of 0.5*M* sodium fluoride calculated on the basis of 3 millimoles of fluoride ion per millimole of cation see Note).

Set up the cell, and ensure that the electrodes are completely covered and the stirring is efficient.

Titrate the solution by adding 0.5*M* sodium hydroxide in 0.1 ml increments and balance the bridge after each addition. If possible, record four or five points on each side of the end point, and then find the latter graphically.

Note:

The amount of fluoride ion is somewhat critical because, if too little is added, a diffuse end point is obtained because of the reaction between the metal ions and the added hydroxide ion; on the other hand, too much fluoride ion causes curvature near the end point, because hydrofluoric acid is a weak acid. The necessary amount of fluoride can be determined to within $\pm$ 25% by trial and error; accurate results are obtained if the amount of fluoride is kept within these limits.

This method is capable of an accuracy of $\pm$ 4% of the amount present with as little as 0.03 millimoles of hydrogen ion; possibly, even better results could be obtained on a larger scale.

(b) Conductometric Titration of Weak Acids and Weak Bases

When weak acids are titrated with strong bases, curves such as 1 and 2 in Fig. IV, 7 are obtained. Before the equivalence point, the conductivity is governed by the extent of ionisation (and by the concentration). The conductivity first decreases as the anion formed suppresses the ionisation, passes through a minimum, and then increases up to the equivalence point because of the conversion of the non-conducting weak acid into its conducting salt. Extensive information about the effect of the dissociation constant on the titration curves of both weak acids and bases has been compiled by Britton[36], and Bruni[37] and Eastman[38] have carried out mathematical analyses of the titration curves of weak electrolytes.

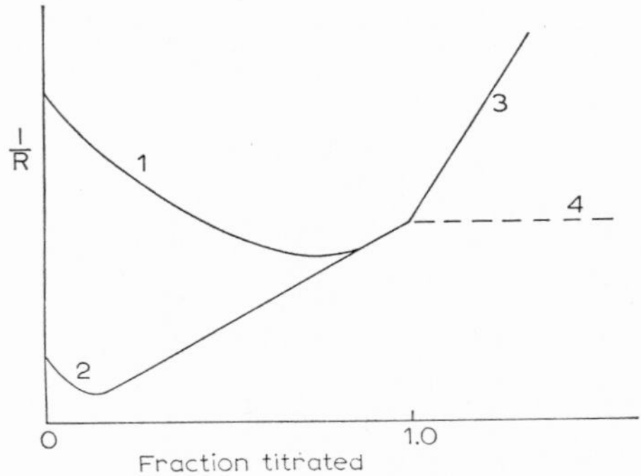

Fig. IV, 7. Conductometric titration of weak acids.
1. $K = 10^{-3}$. 2. $K = 10^{-5}$. 3. Strong base titrant. 4. Weak base titrant.

Near the equivalence point, the graphs of such titrations are often curved because the incompleteness of the reaction allows extra hydroxide ions (or hydrogen ions if a weak base is being titrated) to be present, thus increasing the conductance; the more incomplete the reaction, the more is the curvature.

Conductometric titrations utilise measurements far enough removed from the end point for a fair amount of curvature near the end point to be tolerated. In fact, many titration reactions that are too incomplete for potentiometric end-point detection can be used conductometrically: for instance, boric acid can be titrated conductometrically with sodium hydroxide[39], whereas the pH change at the end point is not large enough for potentiometric or indicator methods to be of use. Weak acids with dissociation constants as small as 10^{-10} can readily be determined, especially in fairly concentrated solution.

(i) Titration of Moderately Strong Acids

Even though the break for the titration of the acid whose dissociation constant is about 10^{-3} (Fig. IV, 7) appears to be well defined, it is often found in practical work that curvature near the end point makes the results uncertain; the titration of 0.01*N* salicylic acid with 0.56*N* sodium hydroxide described by Kolthoff[40] is such a case. In order to find the end point he first plotted the "salt line"; to do this he determined the conductances of the solutions he obtained by the addition to water of a solution of sodium salicylate of the same concentration as that of the alkali; he then took as the end point the point of intersection of the "salt line" and the ordinary titration curve. In this method, the chemical composition of the acid must be

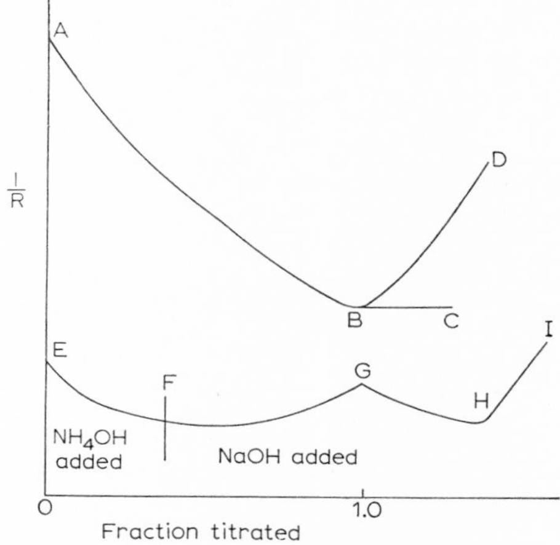

Fig. IV, 8. End-point method of Righellato and Davies[41].

known and any indifferent constituents present in the sample must also be present in the "salt line" determination; for these reasons, this method is not of much practical use.

Righellato and Davies[41] solved the problem by first titrating the moderately strong acid with ammonium hydroxide (line ABC, Fig. IV, 8). If the end point could not be exactly evaluated from this titration, a second was carried out with potassium hydroxide of the same concentration as the ammonium hydroxide (line ABD). Because the potassium ion and the ammonium ion have almost the same equivalent ionic conductances, the curves are iden tical until the equivalence point is reached; a slight correction is usually applied, as the conductivity of the ammonium ion is actually about 0.6% smaller than that of the potassium ion.

References p. 211

A second, more rapid, method was also devised by Righellato and Davies[41], and this is illustrated in the lower portion of Fig. IV, 8. The titration is started with ammonium hydroxide (EF) and is completed with sodium hydroxide of exactly the same concentration (FGHI); EG is the total base required. After G, when all the acid has been neutralised, the conductivity decreases because the ammonium ions are being replaced by the less mobile sodium ions; the final rise (HI) is due to the excess of sodium hydroxide. As the amount of base added between F and H is equal to the amount between E and G, a double check is provided.

(ii) *Titration of Weak Acids*

In the titration of acids whose dissociation constants are 10^{-5} or less, the last part of the neutralisation curve is straight and coincides with the "salt line" (Curve 2, Fig. IV, 7). Generally, there is an advantage in titrating a weak acid with a weak base (Curve 2–4, Fig. IV, 7), rather than with a strong base (Curve 2–3), because the angle between the branches of the curve is usually greater in the former case.

As the dissociation constant of the acid or base titrated becomes smaller, difficulties may be encountered because of hydrolysis, which causes the titration curve to be slightly above the "salt line" and gives rise to an indistinct end point; methods for dealing with this difficulty will be considered in the next section.

In addition to the titration of the many normal weak acids, the conductometric method may be applied to the titration of the hydrogen form of cation-exchange resins[42], for the purpose of determining exchange capacities and equilibrium exchange constants; sodium hydroxide is used as the titrant.

(iii) *Titration of Very Weak Acids and Bases*

As has been mentioned, appreciable hydrolysis occurs near the end point when an acid whose dissociation constant is much less than 10^{-6} is titrated with a strong base. The influence of hydrolysis may be decreased by increasing the concentration of the weak acid, and relatively concentrated solutions should therefore be used. Britton[36] has calculated that it is possible to titrate acids or bases with dissociation constants as small as 10^{-10} because, for $0.1N$ solutions, the first three-quarters of the neutralisation curve will be a straight line; corrections should be made for carbon dioxide and, if this is done, such substances as boric acid and urotropine[39] can be accurately determined.

Because of the difficulty often experienced in recognising the linear ranges of conductometric titration curves when a large amount of hydrolysis occurs, Grunwald[43] has devised a mathematical procedure to replace the ordinary graphical method of locating the end point. This method tends to decrease difficulties due to experimental errors, variation in equivalent conductance, and the dependence of the position of the end point on the selection of straight-line ranges; for 1:1 reactions, a reduction of errors to about 0.5% is claimed. The titration of a very weak base with a strong acid (sodium azide with hydrochloric acid), and the titration of a weak base with a weak acid (ammonium hydroxide with acetic acid), are used as examples. The method, which is given in detail below, is based on the fact that for any two points above $f = 1$ (where f is the fraction titrated) there exist two points below $f = 1$ such that a straight line through the two below will intersect the straight line through the two above at $f = 1$.

(*iv*) *End-Point Calculation Method*

After the conductometric titration curve has been plotted in the usual way, estimate the end point by drawing two straight lines (the usual graphical method). Arbitrarily choose values for f_1, f_2, and f_3 such that f_1 and f_2 are before the end point and f_3 is after the end point. Calculate f_4 using the appropriate formula.

For a weak acid/weak base titration,

$$\frac{f_2 + f_1 - 2f_1f_2}{(1-f_2)\,(1-f_1)} = \frac{f_4 + f_3 - 2}{(f_3-1)\,(f_4-1)}\,.$$

For a strong acid/weak base titration,

$$\frac{1 - f_2f_1}{(1-f_2)\,(1-f_1)} = \frac{f_4 + f_3 - 2}{(f_3-1)\,(f_4-1)}\,.$$

Draw straight lines through f_1 and f_2 and through f_3 and f_4. The intersection of these lines is the new end point. Repeat the process, using the new estimated end point. The new end points found in this way will converge rapidly so long as the first estimated end point was not very greatly in error.

The sharpness of the conductometric end point, and thus the accuracy of the titration of very weak acids, may be improved by first dissolving the sample in an excess of ammonium hydroxide and then titrating with standard lithium hydroxide; this method, which is due to Gaslini and Nahum[44], is based on the fact that the presence of the excess of weak base increases the dissociation of the very weak acid; actually, the ammonium ion is titrated rather than the hydrogen ion. The use of excess of ammonium hydroxide also generally improves the solubility of many samples. The angle at which the branches of the titration curves intersect is much more acute, and the errors for monoprotic acids such as vanillin, phenol, aminobenzoic acid, and guaiacol are less than 1%. The general procedure is as follows:

(v) *Titration of Very Weak Acids*

Apparatus:

For most titrations, a standard Wheatstone bridge and a normal immersion-type of cell are sufficient. The temperature should be controlled to $\pm$ 0.1°. When an accuracy greater than 1% is desired or when compounds that are sensitive to atmospheric oxygen, *e.g.* hydroquinone or pyrogallol, are to be titrated, use a closed cell and a specially constructed bridge (see Note 1) with a Trub-Tauber vibration galvanometer as the zero detector. Carry out the titration in an atmosphere of nitrogen in order to prevent oxidation of the compounds determined.

Procedure:

Add the weighed sample of very weak acid, or an aliquot of its solution (see Note 2) to 120 ml of ammonium hydroxide (see Note 3) in the titration cell. Carry out the titration with an approximately 2.5N solution of carbonate-free lithium hydroxide that has been standardised against potassium biphthalate with phenolphthalein as the indicator. The standard solution can be conveniently delivered from a 5 ml microburette with 0.01 ml graduations.

Notes:

(1) See Reference 18. Undoubtedly, a Jones and Joseph bridge (Leeds and Northrup Co.) would also be an excellent choice.

(2) Choose the sample size so that the sample concentration is about 0.055 equivalents per litre.

(3) The ammonium hydroxide can be obtained free from carbonate by heating the reagent grade concentrated solution in the presence of barium hydroxide and collecting the distillate in carbon-dioxide-free distilled water; its concentration should be about 0.9 equivalents per litre.

(c) *Titration of Acids and Bases Involving more than One End Point*

(i) *Diprotic Acids*

When mixtures of different acids or bases, or polyprotic acids, are titrated, considerable difficulty is sometimes experienced in distinguishing the end point. This is due to the fact that the slopes of the different branches of the titration curve may vary only slightly since, before an excess of hydroxide is present, the differences of slope will depend mainly on the difference in equivalent ionic conductance of the anions involved; Table IV, 2 shows that these differences are apt to be small. In addition, neutralisation of the weaker acid may start before the stronger is completely neutralised. In a few cases, the conductometric titration of mixtures may be more advantageous than potentiometric or indicator methods, but this is not generally so.

Lanzing and van der Wolk[45], who studied the titration of phosphoric acid with sodium hydroxide, found that, as long as the concentration of acid was above about 0.05*M*, a fairly sharp break occurred at the neutralisation of the first hydrogen ion. The first section of the titration curve was similar to the curve for a typical moderately strong acid (see Fig. IV, 9); the subsequent neutralisation points were, however, generally ill-defined and virtually undetectable at concentrations below 0.015*M*.

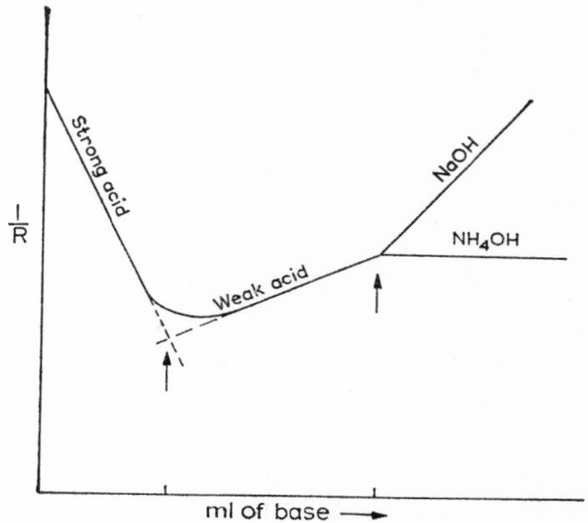

Fig. IV, 9. Titration of a mixture of a strong acid and a weak acid with NaOH or NH_4OH.

Carbonic acid, whose dissociation constants are 3×10^{-6} and 6×10^{-11}, is also characterised by rather poor end points. The neutralisation curve is a combination of that for a weak acid and that for a very weak acid. The neutralisation curve of the first hydrogen ion in H_2CO_3 coincides with the $NaHCO_3$ line. An increase in slope then occurs as HCO_3^-($l_0 = 44$) is converted to CO_3^{2-} ($l_0 = 69$). The slope again increases after the HCO_3^- has been completely neutralised, owing to the addition of excess of hydroxide ions, but this end point is difficult to locate, because the branches of the graph become curved owing to hydrolysis. Kolthoff[46] has recommended that calcium chloride should be added to the solution, and the mixture titrated to the second end point, whose accuracy is greatly improved because of the precipitation of calcium carbonate. The method of Gaslini and Nahum[44] can also be applied to the titration of diprotic acids; these workers successfully titrated resorcinol, hydroquinone and pyrogallol to two end points, and alizarin, phloroglucinol, salicylic acid, *o*-coumaric acid and catechol to a single end point.

References p. 211

(*ii*) *Mixed Acids*

Despite the difficulties sometimes encountered, the titration of mixed acids has received considerable attention. When a strong acid and a moderately strong acid are to be titrated, conductometric titrations are often used, because indicator methods are of no service: for instance, Kolthoff[47] has titrated mixtures of hydrochloric and formic acids with sodium hydroxide; he experienced some difficulty in finding the first end point, because the weaker acid is ionised to an appreciable extent, even though this ionisation is suppressed by the strong acid. Kolthoff obtained the "salt line" by adding a solution of the salt of the strong acid to water; the concentration of the salt solution was equal to that of the sodium hydroxide used for the original titration. The intersection of the salt line and the straight portion of the first, strong acid, branch of the titration curve was taken as the first end point; the second end point was easily found in the usual way.

If the weaker acid has a dissociation constant greater than 3.7×10^{-5}, the concentration being $0.01M$, Kolthoff's method is unsuitable, but Righellato and Davies[41] have devised a rather complex graphical back-titration method to extend its range.

The titration of a moderately strong acid ($K = 10^{-3}$) and a very weak acid ($K = 10^{-10}$) can be carried out very accurately, if both a strong base and a weak base are used as titrants[41]; pyridine is often used as the weak base. The strong base allows the determination of the total acidity but, as the branch of the graph for the moderately strong acid may show curvature, the first end point may not be distinct; the stronger acid can be determined with pyridine or another base that is so weak that it will not react with the very weak acid. Mixtures of mandelic acid and phenol can be determined in this way.

Strong acids can be determined accurately in the presence of weak acids by ordinary methods, so long as the weak acid has a dissociation constant of about 10^{-5} or less. Fig. IV, 9 illustrates such a titration, which finds some industrial application in the titration of strong acids in vinegar.

A rather complex method for the titration of mixtures of two weak acids has been recommended by Schleicher and Pivet[48], who give an empirical formula, the constants of which must be evaluated for each substance titrated.

Conductometric titrations of fatty acids, rosin acids, soaps and their mixtures have been investigated by Mason and his co-workers[49, 50]. In the leather industry, mixtures of sulphuric acid, free sulphonic acids, weak acids and free phenol have been determined in "Syntans"[51].

(d) Replacement Titrations

When a strong acid reacts with the sodium or potassium salts of weak acids, the weakest acid will be replaced first; such a replacement reaction is exemplified by

$$HCl + CH_3COONa = NaCl + CH_3COOH.$$

The investigation of the replacement of a weak acid from its salt was first investigated by Dutoit[52].

Analogous reactions for the replacement of weak bases from their salts by strong bases can, of course, be cited. Kolthoff[53] has studied this type of reaction extensively, and curves for two of Kolthoff's systems are shown in Fig. IV, 10. Curve 1 shows the increase in conductivity during the replace-

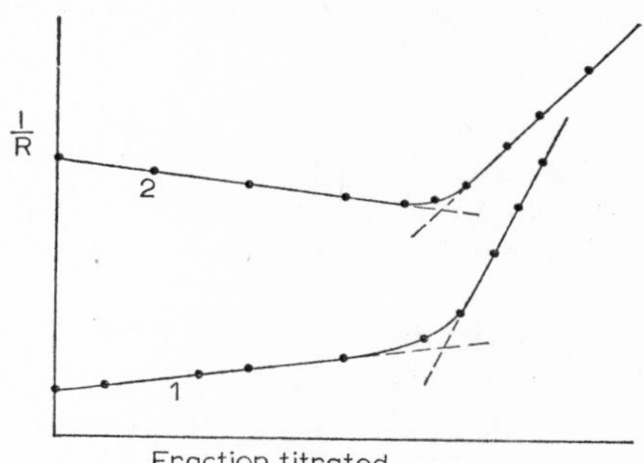

Fig. IV, 10. Replacement conductometric titrations. 1. Sodium acetate with hydrochloric acid. 2. Ammonium chloride with sodium hydroxide.

ment of acetic acid by hydrochloric acid to form sodium chloride; the increase is due to the fact that the mobility of the chloride ion is greater than that of the acetate ion. After the equivalence point, the conductivity increases more rapidly, owing to the addition of excess of hydrochloric acid. Kolthoff has calculated that accurate titrations with $0.01M$ solutions are possible if the dissociation constant of the liberated acid is less than 5×10^{-5}; Lapshin[54] has, however, titrated the salts of weak acids and bases whose dissociation constants are greater than 10^{-4}.

The main application of replacement titrations is in the determination of alkaloids. As alkaloids are weak bases that are often insoluble in water, indicator methods do not work well, and the free base tends to precipitate. Küster, Grüters and Geibel[31] determined quinine, cinchonine, quinidine and cinchonidine by dissolving the alkaloid in hydrochloric acid and back-titrating with barium hydroxide. Dutoit and Levy[55] found that, although direct titrations were not practicable, replacement methods were effective,

References p. 211

provided that the liberated alkaloid base was insoluble; they investigated several reagents that precipitate alkaloids and found that silicotungstic acid yielded the best results, which were accurate to within ± 1% of the alkaloid present. The method is as follows.

Alkaloid Determination with Silicotungstic Acid

Apparatus:

Standard.

Reagents:

Silicotungstic acid. Standardise an 0.05*M* solution of silicotungstic acid gravimetrically (weighing as $SiO_2.12WO_3$) or, preferably, by conductometric titration against a known amount of a pure salt of an alkaloid, such as strychnine.

Procedure:

To an 0.01 to 0.001*N* solution containing *x* g of the alkaloid salt to be determined add 2*x* of sodium acetate. Titrate the mixture with a 0.5 to 0.05*M* solution of silicotungstic acid and plot the titration curve, which can be accurately established with as few as eight points.

Kolthoff[56] was able to titrate a number of alkaloids including tropocaine, atropine, codeine, novacaine and hydrastine, directly with 0.1*N* hydrochloric acid in 50% alcohol solution; using sodium hydroxide as the titrant, he was also able to titrate some commercially available alkaloid salts, such as quinine and cocaine hydrochlorides, by replacement.

(e) Non-aqueous Titrations

The distinction between non-aqueous and aqueous titrations is in some ways vague: for example, phenols are often dissolved in a solvent composed of 40% ethanol and 60% water to enhance solubility and titrated with lithium hydroxide[57]. Generally, however, a non-aqueous solvent is considered to be one in which the amount of water is small or non-existent. Nevertheless, acetone-water mixtures are important enough, and different enough from pure water, for a consideration of their usefulness to be appropriate here.

It has been found that hydrochloric acid can be successfully used as a titrant in water-acetone mixtures containing up to 80% of acetone[58]; above 80%, the conductivity of sodium chloride increases, and that of hydrochloric acid decreases, to such an extent that no break occurs. Alcoholic sodium hydroxide can also be used as a titrant for acetone-water mixtures. It has been found that the behavior of numerous acids depends on the amount of acetone present: in 80% acetone : 20% water, sulphuric, sulphurous, picric, and benzenesulphonic acids behave as strong (and monoprotic) acids; in 40% acetone : 60% water, sulphuric acid behaves as a diprotic acid.

Cheshire, Brown, and Holms[59] have developed a procedure for the titration of acid mixtures based on the behaviour of acids in acetone mixtures: any bases are first removed from the sample by passing it through an anion exchange resin; three titrations with sodium hydroxide are then carried out, the first in 80% acetone, the second in 40% acetone, and the third in water. Acids of pK less than 1 are determined by the first titration; the difference between the second and first titrations gives the acids with pK's between 1 and 2, and the difference between the third and the second gives those acids of pK greater than 2. The interference of phenolic acids is eliminated by adding excess of pyridine, when the difference between the third and second titration gives acids of pK 2–7. The salt content is obtained from the difference between the titres before and after the bases have been removed.

A number of completely non-aqueous conductometric titrations have also been developed, although Hall and Spengeman[60] have concluded that conductometric end-point detection is inferior to potentiometric methods in the titration of organic bases. Higuchi and Rehm[61] have used anhydrous acetic acid as a medium for titrating mixtures of sulphuric and hydrochloric acids with lithium acetate in acetic acid; three breaks were obtained, one for hydrochloric and two for sulphuric acid. The first hydrogen ion of the sulphuric acid is titrated first, then the hydrochloric acid and finally the second hydrogen ion of the sulphuric acid. Sulphuric acid has also been titrated in glycol or glycol-acetone with piperidine[62].

McCurdy and Galt[63] have titrated mixtures of a number of weak bases in a 1:1 mixture of 1,4-dioxan and formic acid, using perchloric acid in the same solvent as a titrant. Pinketon and Briscoe[64] have titrated weak acids such as *m*-chlorobenzoic, benzoic, sulphanilic, *p*-aminobenzoic, and α-aminobutyric acids in monoethanolamine; the sodium salt of monoethanolamine was used as the titrant.

Weak carboxylic and phenolic acids have been titrated in various solvents such as methanol, dimethylformamide, and pyridine[65]; potassium methoxide in pyridine-benzene-methanol and tetramethylammonium hydroxide in benzene were both used as titrants. Precipitation of salts was often encountered, but the results for aromatic dicarboxylic acids were stated to be accurate to ± 1%, and for alkylphenols to ± 2%.

Even fused salts can be used as media for conductometric titrations, as is shown by the work of Jander and Brodersen[66], who carried out acid-base titrations in fused mercuric bromide, using mercuric sulphate as the acid and potassium bromide as the base.

References p. 211

(f) Titration of Complex and Heteropoly Acids

Molybdic and tungstic acids have been studied by Britton and German[67] and by numerous other workers[69, 70], and the existence of a number of complex species is indicated. Molybdenum and tungsten have been determined by Rother and Jander[68], who titrated solutions of sodium molybdate with lead acetate and those of sodium tungstate with lead nitrate; these methods are said to be more rapid and accurate than the classical gravimetric methods.

4. Precipitation Titrations

(a) Theoretical Considerations

Precipitation titrations can be effectively followed by conductometric methods, though not so well as acid-base titrations, which are characterised by sharp breaks because both the hydrogen ion and the hydroxyl ion have very high equivalent ionic conductances. In precipitation titrations, one pair of ions is substituted for another and, as the experimenter has a choice of reagents, good results can usually be obtained. If a cation is to be precipitated, a titrant whose cation has the smallest possible mobility is selected and, if an anion is to be precipitated, a titrant whose anion has as small a mobility as possible; in this way, the maximum possible change in conductance during a titration is assured.

Errors associated with conductometric precipitation titrations can usually be traced to one of three causes: fouling of the electrodes, adsorption and co-precipitation, incomplete precipitation. The electrodes used to determine the conductance often become fouled, or contaminated with the precipitate, and this causes erratic readings; this difficulty can be minimised by mounting the electrodes vertically and by titrating at an elevated temperature that is controlled to within at least 1°; the use of high frequency methods is also effective because the electrodes do not normally come in contact with the solution (see Chapter V). Adsorption and other forms of co-precipitation can cause interference, just as in any precipitation method; effective stirring is usually the best answer to this trouble, but very often extraneous materials that are preferentially adsorbed are added to the titration solution. Incomplete precipitation causes indistinct end points owing to excessive curvature of the various branches of the titration curve. If incompleteness of precipitation is due to the formation of supersaturated solutions, titration at an elevated temperature, or seeding with small crystals of the substance precipitated, is often effective. If excessive solubility is the problem, the addition of alcohol or acetone to the titration medium often produces accurate results. If difficulties due to supersaturation are negligible or can be avoided, the classical solubility product may be used to describe the titration curve with reasonable accuracy, provided that the concentration of electrolyte in the solution to be titrated is known; in any case, the contribution of the slightly soluble precipitate to the solubility can be calculated.

The solubility product of a slightly soluble salt can be determined by the measurement of the conductivity (resistance) of a saturated solution in the pure solvent:

$$C = \frac{1000\,\theta}{(l_0{}^+ + l_0{}^-)\,R}, \tag{20}$$

where C is the concentration of the salt, R and θ have the significance indicated on pp. 174, 175 and $l_0{}^+$ and $l_0{}^-$ are the equivalent ionic conductivities of the cation and the anion respectively.

For instance, the specific conductance of a saturated solution of silver chloride was found[71] to be 1.8×10^{-6} mhos after correction for the conductivity of the pure water used; the equivalent ionic conductance of the ions involved can be found from the appropriate table (see Table IV, 2), so that the concentration C may be calculated as follows:

$$C = \frac{1000 \times 1.8 \times 10^{-6}}{62 + 76} = 1.3 \times 10^{-5} \tag{21}$$

The concentration C refers to the concentration of silver chloride in solution and also to the concentration of both chloride ions and silver ions.

The solubility product S can then be computed:

$$\begin{aligned} S &= [Ag^+]\,[Cl^-] = 1.3 \times 10^{-5} \times 1.3 \times 10^{-5} \\ &= 1.7 \times 10^{-10} \end{aligned}$$

Not only does the magnitude of the solubility product of a slightly soluble salt govern the equilibrium between the solid and the liquid phase, but the solubility product must, of course, be exceeded for precipitation to occur at all. Table IV, 3 lists the solubility products of a number of precipitates that often occur in conductometric titrations.

Precipitation titrations are actually made possible, regardless of what is being titrated or what titrant is selected, because, so long as a precipitate is forming, the conductivity of the solution will change only slightly. The best breaks are obtained if, as previously mentioned, the conductivity decreases slightly up to the equivalence point; after the equivalence point, the conductivity always rises, because an excess of electrolyte (the titrant) is being added.

References p. 211

TABLE IV, 3

SOLUBILITY PRODUCTS AT 25°

Substance	*Solubility Product*	*Substance*	*Solubility Product*
AgCl	1.7×10^{-10}	$BaSO_4$	1.1×10^{-10}
AgBr	7.7×10^{-13}	$BaCrO_4$	2×10^{-10}
AgI	1.5×10^{-16}	$BaCO_3$	8.1×10^{-9}
Ag_2CrO_4	9×10^{-12}		
AgSCN	1.2×10^{-12}	$CaCO_3$	8.7×10^{-8}
		CaC_2O_4	2.6×10^{-9}
PbI_2	1.4×10^{-8}	CaF_2	3.9×10^{-11}
PbF_2	3.6×10^{-8}		
PbC_2O_4	2.7×10^{-11}	CuSCN	1.6×10^{-11} at 18°
$PbSO_4$	1.1×10^{-8}	CuI	5.1×10^{-12} at 18°
$PbCrO_4$	1.8×10^{-14}		

The concentrations of the solutions being titrated, and consequently that of the titrant, are important variables; in general, the selection of dilute solutions is desirable for the elimination of co-precipitation; if, however, solutions are too dilute, the amount of precipitate going into solution will increase correspondingly, and excessive curvature of the titration curve will result.

Kolthoff[72] considers that satisfactory conductometric titrations can be made if the first half of the precipitation section of the graph is straight. (A contribution to the total conductivity of 1% from the dissolution of the precipitate is considered to constitute curvature.) On this basis, it can be calculated that titrations involving precipitates whose solubility products are less than 5×10^{-5} can be carried out at a concentration of $0.1N$, and that, if the concentration to be used is reduced by 10^{-1}, the titrant must be so chosen that the solubility product of the resulting precipitate is 10^{-2} less. The work of Freak[73] on the precipitation of barium sulphate and silver chloride tends to confirm Kolthoff's conclusions.

Britton[74] has derived the following formula for the angle ψ between the branches of the titration curve:

$$\tan \psi = \frac{\dfrac{N}{1000x}\begin{pmatrix}\textit{sum of the mobilities}\\ \textit{of ions of precipitate}\end{pmatrix}}{1 + \left(\dfrac{N}{1000x}\right)^2 \begin{pmatrix}\textit{mobility of cation}\\ \textit{(anion) in precipitant}\\ -\ \textit{mobility of cation}\\ \textit{(anion) in precipitate}\end{pmatrix}\begin{pmatrix}\textit{sum of ionic}\\ \textit{mobilities of}\\ \textit{precipitant}\end{pmatrix}}, \tag{22}$$

where x is the volume in ml of titrant required to reach the end point, and N is the normality of the solution titrated.

Fig. IV, 11 shows some theoretical titration curves illustrating the importance of ψ and of its magnitude. For the specific example illustrated in curve 1, equation (22) becomes

$$\tan\psi = \frac{\dfrac{N}{1000x}(l_{\mathrm{Ba}^{2+}} + l_{\mathrm{SO_4}^{2-}})}{1 + \left(\dfrac{N}{1000x}\right)^2 (l_{\mathrm{K}^+} - l_{\mathrm{Ba}^{2+}})\ (l_{\mathrm{K}^+} + l_{\mathrm{SO_4}^{2-}})} \tag{23}$$

The sharpest and most useful titration curves will be characterized by small values of ψ, which can be achieved by making the denominators in the above equations as small as possible: if the equivalent ionic conductance of the cation of the precipitant is made as small as possible, the second term

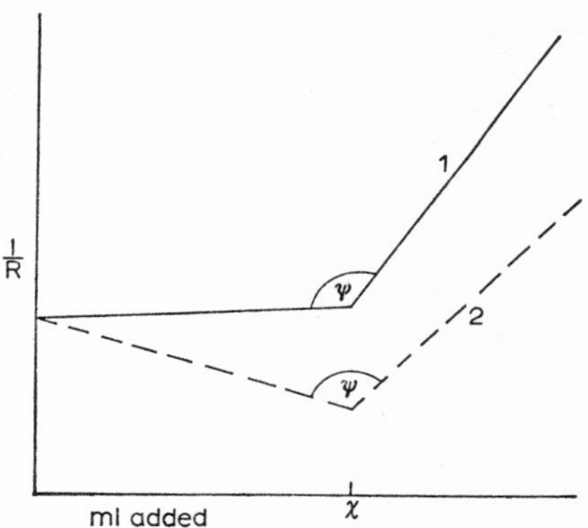

Fig. IV, 11. Conductometric titrations of barium chloride with: 1. potassium sulphate; 2. lithium sulphate.

of the denominator will be negative and a small value for ψ will result. Minimum values for ψ can also be attained by making x as small as possible, in other words, by making the titrant as concentrated as is practicable. Equation (22) shows also that the sharpness of the angle at the end point will be greater, if the conductance of the titrant ion that enters the precipitate is as large as possible; thus, better results are obtained in the titration of barium, if sulphate (l_0 = 80) is used rather than carbonate (l_0 = 69).

In the derivation of equations (22) and (23) and in the construction of Fig. IV, 11 it has been assumed that the precipitates in question are so insoluble that there is little or no distortion of the titration curve; in actual cases, the solubility product must be considered as well as the mobilities of the various ions involved.

References p. 211

Some efforts have been made to eliminate the errors that are due to adsorption, slowness of precipitation and electrode fouling, and are common to a number of conductometric precipitation titrations: carrying out the titrations at elevated temperature[75, 76], *e.g.* in boiling solution, apparently gives accurate results, but there is insufficient evidence to judge these methods critically.

There has been considerable discussion in the literature about the accuracy of several conductometric titrations, especially the determination of sulphate by titration with barium[77–80]; many cations and anions, such as aluminum, zinc, calcium, acetate, tartrate and malate have been said to cause errors. It may be concluded, however, that alkali sulphates can be determined with an accuracy of $\pm$ 1% of the amount present, when barium chloride is used as the titrant.

The formation of unknown or unexpected precipitates such as basic salts interfere with many other titrations. The careful measurement of specific conductivities often gives the experimenter an insight into exactly how the precipitation is proceeding, and such measurements are recommended during the development of new methods, as well as the establishment of procedures that give sharp end points.

(b) Reagents and Procedures

A number of precipitation titrations are recorded in Table IV, 4. Brief remarks are included to aid the experimenter but, in many cases, more extensive consideration is desirable.

(*i*) *Silver Nitrate*

The reaction of the silver ion with the anions listed is generally straightforward, but if silver nitrate is added to an alkali cyanide solution a small break is observed when one-half of the theoretical amount has been added; this results from the formation of the soluble silver cyanide complex,

$$2\,CN^- + Ag^+ = Ag(CN)_2^-;$$

further addition of silver ion causes a precipitate to form,

$$Ag^+ + Ag(CN)_2^- = Ag\,[Ag(CN)_2],$$

and the completion of the latter reaction is marked by a well defined break. As the instability constant of the silver cyanide complex is about 10^{-22}, cyanide may be determined in the presence of halides, and Kolthoff[81] has determined cyanide-chloride mixtures conductometrically with good results.

Silver halides dissolve in ammonium hydroxide to an extent that depends

TABLE IV, 4

CONDUCTOMETRIC PRECIPITATION TITRATIONS

Titrant	*Substances Determined*	*Remarks*	*Ref.*
Silver nitrate	Chloride		81, 82
	Bromide		81
	Iodide		81
	Thiocyanate		81
	Cyanide		81
	Chromate		81
	Ferricyanide		81
	Selenite	pH = 9	83
	Cyanate		84, 86
	Selenocyanide		85
	Molybdate		87
	Tungstate		87
Lead nitrate	Iodide	At least 0.05M	88
	Ferrocyanide		88
	Sulphate	Alcohol added to reduce solubility	88, 32
	Sulphite		88
	Thiosulphate	At least 0.05M	88
	Pyrophosphate		88
	Molybdate	Dissolve in alkali, neutralise to phenolphthalein with acetic acid	68, 89
	Tungstate	Dissolve in alkali, neutralise to phenolphthalein with acetic acid	68
	Selenite		90
	Selenate		91
	Oxalate	Neutral solution	88
	Tartrate	Neutral solution	88
	Succinate	Neutral solution	88
	Benzoate		88
Barium acetate (or chloride)	Sulphate		79, 52, 80
	Chromate		79
	Carbonate	0.01M; add alcohol if solution dilute	79
	Pyrophosphate		79
	Oxalate	50% alcohol	79
	Tartrate	50% alcohol	79
	Citrate	50% alcohol	79
	Selenate		91
Thallous sulphate	Iodide		92
Lithium sulphate	Barium		92, 96
	Strontium	Best results with solutions about 0.002M with the addition of alcohol (Strontium, Calcium, Lead)	92, 96
	Calcium		92, 96
	Lead		92, 96
Uranyl acetate (or nitrate)	Phosphate		52, 93, 94, 95, 73
	Arsenate	Readings not always constant. Titration should be done rapidly.	52, 95
Sodium (or lithium) chromate	Barium		92, 98
	Strontium	40–50% alcohol necessary	92, 98

References p. 211

TABLE IV, 4 (*continued*)

Titrant	*Substances Determined*	*Remarks*	*Ref.*
Sodium (or lithium) chromate (*continued*)	Lead		52, 98
	Silver		98
	Thallium(I)		98
Sodium sulphide	Zinc	Acetic acid	99
	Copper	3% accuracy	99
	Bismuth	Nitric acid	99
	Silver	3% accuracy	99
	Cadmium		99
	Ferrous iron		99
Hydrogen sulphide	Lead		100
	Cadmium		100
	Copper		100
	Silver		101
	Bismuth		100
Lithium oxalate	Calcium		92, 95, 73, 97
	Barium		97
	Strontium		97
	Silver		97
	Lead		97
	Copper		97
	Cadmium		97
	Nickel	Alcohol and concentrated solutions necessary	97
	Cobalt		97
	Manganese		97
	Ferrous iron	Results not satisfactory	97
	Magnesium		97
Potassium ferrocyanide	Zinc	Composition of other metal ferrocyanides varies according to conditions	101
	Lead		101
Calcium ferrocyanide	Potassium	36% ethanol $K_2CaFe(CN)_6$	102
Potassium ferricyanide	Cobalt		101
	Nickel		101
	Cadmium		101
	Copper		101
Lithium halides	Silver		52, 96, 100
Bismuth oxyperchlorate	Phosphate		103
Sodium perchlorate	Potassium	Not accurate	104
8-quinolinol	Copper		105
Barium hydroxide	Zinc (sulphate)	$KMnO_4$ reduces co-precipitation	106
	Magnesium (sulphate)	In presence of calcium	76
	Cobalt (sulphate)		76
	Nickel (sulphate)		76
	Calcium bisulphite and sulphurous acid	Oxidise with H_2O_2	107

on their solubility products and the stability constants of their ammonium complexes. Dutoit and Reed[108] were able to titrate mixtures of chlorides and bromides by first determining the total halide by titration of a neutral or slightly acidic solution with silver nitrate; the bromide was then determined in ammoniacal solution and the chloride found by difference. As the amount of ammonium hydroxide added is critical, and depends on the amount of chloride and bromide present, the method is not practical. Kolthoff[81] has shown that iodide can be determined in the presence of chloride and less accurately in the presence of bromide; he used $0.1N$ solutions of halide and titrated them with $1N$ silver nitrate, adding from 6 to 25 ml of 10% ammonium hydroxide to mask the chloride and bromide.

(*ii*) *Lead Nitrate*

Lead nitrate can be used for the determination of ferrocyanide, because a definite salt, $Pb_2Fe(CN)_6$, is formed[88] and, as lead does not form a precipitate with ferricyanide, this method has considerable practical importance.

Thiosulphate can be accurately determined in $0.05M$ solution with lead nitrate, even though the precipitate is so soluble that it redissolves until more than one-fifth of the stoichiometric amount has been added. After a permanent precipitate has been obtained, the conductivity remains constant until excess of lead nitrate is added. Possibly, the success of this titration depends upon lead thiosulphate being undissociated even though it seems to be appreciably soluble.

(*iii*) *Barium Acetate (or Chloride)*

Barium acetate is especially useful for the determination of sulphate, and Jander and his co-workers[109] have devised the following procedure for the determination of sulphate in drinking water.

Determination of Sulphate in Drinking Water

Procedure:

Decompose the bicarbonates in a 50 ml sample of water by boiling, cool the water quickly and filter off any calcium carbonate. Add 50 ml of ethanol and titrate the water rapidly (Note 1) with barium acetate at room temperature, recording and plotting the conductance in the usual way. If desirable, separate samples may be taken for the determination of chloride by titration with silver nitrate, and of the total permanent hardness (Note 2) by titration with $0.1N$ lithium oxalate in 30–40% ethanol.

Notes:

(1) A rapid titration is necessary to avoid the precipitation of calcium sulphate.

(2) Provided that the calcium: magnesium ratio is greater than 10:1.

References p. 211

(iv) *Lithium Oxalate*

Lithium oxalate has been used particularly for the determination of calcium[92, 95, 73, 97], and Kolthoff[97] has also tried to use it with a number of univalent and bivalent metal ions; as, however, oxalate tends to form stable complexes with most heavy metals, practical titrations were not possible. Titration of barium and strontium is improved by the addition of alcohol, although strontium gives good results without alcohol, except at low concentrations.

When lead or copper is titrated in approximately 0.1*M* solution, the precipitate begins to dissolve before the end point is reached, because of complex formation; as a result, the titration curve is composed of two sections, but a sharp break occurs at the end point, allowing an accuracy of $\pm$ 1% of the amount present. Cadmium behaves similarly, but the results are only accurate to $\pm$ 2% of the amount present; nickel and cobalt give satisfactory results only in the presence of alcohol; with manganese and ferrous iron, unsatisfactory results are caused by excessive complex formation.

(v) *Mercuric Perchlorate*

There are a number of mercuric salts that, while soluble in aqueous solution, are so completely undissociated that reactions that yield them can be used for conductometric titrations[110, 111]. Kolthoff[110] has shown that mercuric perchlorate is a suitable titrant for chloride, bromide, iodide, cyanide, thiocyanate, formate, acetate and its homologues; oxalate, tartrate, and citrate, on the other hand, give unsuitable curves. Kolthoff applied this method to the determination of chloride in drinking water with an error of only $\pm$ 2–3% of the amount present at a concentration as low as 0.00025*M*; small amounts of sulphate do not interfere, but bicarbonate causes an error that must be eliminated by the addition of nitric acid.

Solutions of mercuric perchlorate can be made by treating an excess of red mercuric oxide with 2*M* perchloric acid.

(vi) *Sodium Polysulphide*

The precipitation of sulphides with sodium polysulphide, Na_2S_9, has been studied[99], but the possibility that the reagent may be contaminated by species such as thiosulphate throws some doubts on its usefulness. Although sulphides are among the most insoluble of precipitates, difficulties in the preparation of reagents and lack of uniformity in the composition of the precipitates have caused much trouble with attempted applications.

(vii) *Sodium Hydroxide and Barium Hydroxide*

In general, conductometric titrations involving the precipitation of metallic bases are of little use analytically, because the hydroxides are often precipitated in an impure state. The reverse of the normal titration procedure, *i.e.* the addition of metal ions to an excess of hydroxide, is said to yield good analytical results[112], but has not been extensively applied.

Much useful information can be gained from conductometric studies of titrations with sodium hydroxide, as may be seen from the work of Robinson and Britton[113], who titrated aluminium sulphate with sodium hydroxide, and showed the existence of several basic aluminium sulphates as well as of sodium aluminate. (For an extensive discussion see Reference 1, Chapter 13.)

Harned[76] has recommended barium hydroxide as a titrant for bivalent metal sulphates, because two precipitates are formed

$$Ba(OH)_2 + MSO_4 = \underline{BaSO_4} + \underline{M(OH)_2},$$

causing a sharp break in the titration.

Magnesium can be determined by Harned's method (calcium does not interfere, because calcium hydroxide is soluble): carbon dioxide is first removed by acidifying the sample solution with sulphuric acid and boiling it; the excess of sulphuric acid is neutralised to the phenolphthalein end point with 0.1*M* barium hydroxide, which is then used to titrate the metal; magnesium in dolomite can be determined to within $\pm$ 1 mg in this way.

Copper, cobalt and nickel can also be determined with barium hydroxide, but the titration is best done hot to prevent the formation of oxysulphates; with cobalt and nickel there were some difficulties due to the excess of sulphuric acid. Cadmium cannot be determined accurately, and the precipitation of $CdSO_4 \cdot 3Cd(OH)_2$ causes the titration curve to show two breaks.

Jander and Jahr[107] have used barium hydroxide to titrate mixtures of sulphurous acid and calcium bisulphite, the sulphite ions being first oxidised with H_2O_2. A titration curve with two breaks is obtained: the first break indicates complete neutralisation of the sulphuric acid, and the second corresponds to the completion of the reaction

$$CaSO_4 + Ba(OH)_2 = Ca(OH)_2 + \underline{BaSO_4}.$$

(*viii*) *Lithium or Sodium Tetraphenylboron*

The conductometric titration of potassium in dilute acetic acid with lithium tetraphenylboron has been found by Raff and Brotz[114] to give results that compare favourably with those of gravimetric methods, and Wendlandt[115] has reported the conductometric precipitation titration of thallium with sodium tetraphenylboron; interferences were not investigated, but ions such as Hg^{2+}, Hg_2^{2+}, K^+ and Ag^+, which form precipitates with tetraphenylboron, should be absent.

References p. 211

(ix) *Analysis of Rubber Chemicals*

Several conductometric titration procedures for the analysis of sulphur in rubber and rubber chemicals[116, 117, 118] have been described. The standard method[116] consists in burning a sample containing 5–6 mg of sulphur in oxygen and absorbing the resulting sulphur dioxide in neutral hydrogen peroxide; the solution is neutralised with ammonium hydroxide, the excess of peroxide is destroyed with manganese dioxide, and the solution is evaporated to dryness; the residue is dissolved in water and the solution titrated conductometrically with barium acetate.

A more elegant method[53] is to treat an alcoholic solution of the sample (for instance free sulphur from vulcanisates) with excess of potassium cyanide for eight hours at 80°; nitric acid is added and the excess of hydrogen cyanide is driven off by boiling; the thiocyanate formed is finally titrated conductometrically with silver nitrate.

5. Complexometric Titrations

Conductivity has long been used in the study of complex formation; Werner, for example, supported many of the postulates of his co-ordination theory with conductivity measurements. Under controlled conditions, conductivity measurements can be used to show what type of electrolyte, *e.g.* uni-univalent, a complex salt gives in solution. (For an excellent discussion of Werner's work and that of his critics see reference 119.)

Conductometric titrations have also been applied to the study of complex salts: for instance, Job[120] titrated roseo-cobaltic sulphate $[Co(NH_3)_5H_2O]_2(SO_4)_3$ with barium hydroxide, finding two breaks in the titration curve corresponding to the following reactions

$$[Co(NH_3)_5H_2O]_2(SO_4)_3 + Ba(OH)_2 = [Co(NH_3)_5OH]_2(SO_4)_2 + BaSO_4,$$
$$[Co(NH_3)_5OH]_2(SO_4)_2 + 2Ba(OH)_2 = [Co(NH_3)_5OH](OH)_4 + 2BaSO_4,$$

and thus confirming the structure suggested by Werner.

Dey[121] has used conductometric titrations to confirm the existence of a number of copper(II)-ammine complexes: by plotting the deviations from the sum of the conductivities of the constituents against composition, he found maxima corresponding to three, four, five and six moles of co-ordinated ammonia per mole of copper.

Conductometric titration of the thiosulphate complexes of thallium[122] showed that $Tl_2S_2O_3$, $Tl_2(S_2O_3)_2^{2-}$ and $Tl(S_2O_3)_2^{3-}$ exist in solution. Similarly, a study of the complex carbonates of uranium[123] has given evidence for the existence of $UO_2(CO_3)_3^{4-}$ and $UO_2(CO_3)_2^{2-}$.

The reactions of lanthanum and thorium with tartrate and hydroxide have been extensively studied by Britton and Battrick[124], who found indications of various basic tartrate complexes.

The complexes of chromium have been extensively investigated by Shuttleworth, whose work is of special interest, not only because of its importance to the leather chemist, but also because chromium forms such a wide variety of complexes, which often involve OH bridges; the complexes of chromium with chloride[125], sulphate[126], lactate[127], monoprotic organic acids[128] and polyprotic organic acids[129], have all been studied by Shuttleworth.

Theis and Serfass[130] found that conductometric titrations are especially suitable for the determination of "olation" (linkage between complex metal ions by bridging OH groups). Similarly, the effect of aging on the olation in chrome tanning liquors has been studied[131] by the titration of 0.06*N* chrome solutions with 1*N* hydrochloric acid, and it was found that six days at 40° were necessary for olation equilibrium to be established.

Britton and Dodd[132] have studied the formation of the cyanide complexes of cadmium, silver, nickel, zinc and mercury by conductivity methods; a titration procedure was not used, because of the precipitates formed with all these metals except mercury. Instead, various amounts of 0.1*M* potassium cyanide were added to 10 ml of a 0.1*M* solution of the metal salt, and the conductivity was measured after time had been allowed for equilibrium to be attained. No well defined "end points" for the formation of the cyanide complexes were found, except with silver, but the existence of complexes was demonstrated. The cyanide complexes of mercury have also been studied by Gaugin[133].

(a) EDTA Methods

Since 1948, complexometric titrations have become of great analytical usefulness, because of the value of ethylenediaminetetra-acetic acid (EDTA) as a titrant (see Vol. 1B, p. 288), but only a few papers have appeared in which conductance is used as a means of detecting the end point; this state of affairs is entirely reasonable in view of the many excellent indicators that have been suggested.

Hall *et al.*[134], who first evaluated conductometric titrations for the standardisation of solutions of bivalent metal ions, obtained excellent accuracy, but noted that conductometric methods gave lower molarities than indicator methods; this was attributed to the partial dissociation of the complex near the end point causing the indicator to change colour before the equivalence point, but it is more likely that these errors are associated with incorrect interpretation of the colour changes involved.

Sharper end points were obtained with more dilute solutions of the ions being determined; this effect is partly attributed to the fact that EDTA complexes are more stable at lower ionic strengths. Actually, the sharper end point seems to be associated with the fact that there is a greater percentage change in conductivity when a titrant of a given strength is added to a dilute solution than to a more concentrated one.

Hall and his co-workers[134] also showed that the addition of a buffer solution is necessary to obtain satisfactory end points; this is usually the case with EDTA titrations, but is especially important with conductometric end-point detection, because the hydrogen ion has such a high mobility. EDTA is usually added in the form of the disodium salt, so that hydrogen ions are released as the metal ions are complexed

$$M^{2+} + Na_2H_2Y = 2Na^+ + MY^{2-} + 2H^+.$$

If this liberated hydrogen ion has no buffer with which to react, the conductivity increases rapidly from the start of the titration, and the end point is obscured. The minimum amount of buffer is, however, desirable in order to keep the concentration of the excess of electrolyte as low as possible; this is another reason for carrying out conductometric titrations with EDTA at as low a concentration as possible.

Vydra and Karlik[135] have also shown that conductometric methods of end-point detection can be applied to chelometry; in addition, they have determined the total hardness of coloured and turbid waters, in which visual indicators could not be used. Oxidising agents sometimes react with indicators such as Eriochrome Black T but do not affect conductometric end points. Vydra and Karlik's method is as follows.

Total Water Hardness by Conductometric Titration with EDTA

Procedure:

To a 200 ml sample of water add 5 ml of a 0.05*M* solution of sodium borate, $Na_2B_4O_7$, 0.3 ml of a 10% solution of potassium cyanide and 0.3 ml of a 10% solution of triethanolamine (see Note). Titrate the mixture with a 0.02*M* solution of the disodium salt of EDTA, measuring the conductivity after each addition of the titrant. Plot the titration curve and determine the end point graphically.

Note:

The addition of these reagents not only serves to buffer the solution, but also masks any heavy metals that would ordinarily interfere.

Martell[136] has discussed the titration of EDTA with sodium hydroxide and with calcium hydroxide, illustrating both the weak acid and the complexing properties of EDTA.

(b) Conductometric Determination of Fluoride

Harms and Jander[137] have described a volumetric determination of fluoride in which they used changes in conductance to locate the end point. The titrant was aluminium chloride and the sample was diluted with dilute acetic acid. For milligram quantities, the method is accurate to $\pm$ 1% of the amount present. Kubota and Surak[138] have used lanthanum acetate for the same determination, and their method is as follows.

Determination of Fluoride

Reagent:

Lanthanum acetate. Dissolve 0.1 mole of pure lanthanum nitrate in water and precipitate the lanthanum with 30 ml of ammonium hydroxide sp. gr. 0.880. Filter and wash the precipitate with water and dissolve it in 25 ml of glacial acetic acid. Dilute the solution to one litre and standardise it by evaporating a 25.00 ml portion to dryness and igniting the residue to a constant weight of lanthanum oxide at 950°.

Procedure:

To a 5 ml sample (see Note 1) containing about 1 mg of fluoride add 50 ml of water and 10 ml of ethanol. Titrate the fluoride conductometrically with 0.1*M* lanthanum acetate delivered from a microburette, and plot the conductivity readings against the volume of titrant and locate the end point graphically (see Note 2).

Notes:

(1) Aliquots of a distillate may be used, if it is necessary to separate the fluoride from interfering substances in the sample.

(2) There may be some curvature near the end point, but good results can be obtained because the conductivity readings on either side of the end point should fall on straight lines.

This method is quite useful, because the partial dissociation of the lanthanum fluoride complex makes it necessary to use some extrapolation procedure to detect the end point.

References

1. H. C. Parker and E. W. Parker, *J. Amer. Chem. Soc.*, 1924, **46**, 312.
2. G. Jones and B. C. Bradshaw, *J. Amer. Chem. Soc.*, 1933, **55**, 1780.
3. B. E. Conway, *Electrochemical Data* (Elsevier Publishing Co., Amsterdam, 1950), p. 142.
4. D. A. MacInnes, *The Principles of Electrochemistry* (Reinhold Publishing Corporation, New York, 1939), p. 342.
5. R. F. Hall and L. H. Adams, *J. Amer. Chem. Soc.*, 1919, **41**, 1515.

6. M. Randall and G. N. Scott, *J. Amer. Chem. Soc.*, 1927, **49**, 636.
7. G. Jones and R. C. Joseph, *J. Amer. Chem. Soc.*, 1928, **50**, 1049.
8. T. Yoshida and H. Hattori, *J. Chem. Soc. Japan, Ind. Chem. Sect.*, 1950, **53**, 47, 103.
9. J. T. Stock, *Analyst*, 1948, **73**, 600.
10. M. Buras and J. D. Reid, *Ind. Eng. Chem., Analyt.*, 1944, **16**, 591.
11. G. Jander and H. Schorstein, *Z. angew. Chem.*, 1932, **45**, 701.
12. C. Morton, *Trans. Faraday Soc.*, 1937, **33**, 474.
13. L. J. Anderson and R. R. Ravelle, *Analyt. Chem.*, 1947, **19**, 264.
14. M. I. Kulenok, *Zavodskaya Lab.*, 1955, **21**, 1027.
15. A. Juliard and J. van Cakenberghe, *Analyt. Chim. Acta*, 1948, **2**, 542.
16. R. D. Goodwin, *Analyt. Chem.*, 1953, **25**, 263.
17. H. T. S. Britton and L. W. German, *J. Chem. Soc.*, 1930, 1250.
18. N. Kano, *J. Chem. Soc. Japan*, 1922, **43**, 556.
19. N. Ya. Khlopin, *Zavodskaya Lab.*, 1940, **9**, 962.
20. R. H. Müller and A. M. Vogel, *Analyt. Chem.*, 1952, **24**, 1590.
21. E. D. Eastman, *J. Amer. Chem. Soc.*, 1920, **42**, 1648.
22. R. P. Taylor and N. H. Furman, *Analyt. Chem.*, 1952, **24**, 1931.
23. R. Duval and P. Duval, *Compt. rend.*, 1937, **205**, 1237.
24. M. Duval and C. Duval, *Compt. rend.*, 1945, **220**, 115.
25. P. Delahay, *Analyt. Chem.*, 1948, **20**, 1215.
26. F. P. Treadwell, *Helv. Chim. Acta*, 1920, **1**, 97.
27. I. M. Kolthoff, *Bull. Soc. chim. France*, 1910, **7**, 1.
28. P. Dutoit, *J. Chim. phys.*, 1910, **8**, 12.
29. P. Dutoit and P. B. Mojoiu, *ibid.*, 27.
30. W. Poethke, *Z. analyt. Chem.*, 1931, **86**, 45.
31. F. W. Küster, M. Grüters, and W. Geibel, *Z. anorg. Chem.*, 1904, **42**, 225.
32. E. Müller and H. Kogert, *Z. anorg. Chem.*, 1930, **188**, 60.
33. J. D. Clark, *Naval Air Rocket Test Sta. Rept., 1951*, No. 9.
34. I. G. Murgulescu and E. Latiu, *Z. analyt. Chem.*, 1948, **128**, 142.
35. L. P. Pepkowitz, W. W. Sabol, and D. Dutina, *Analyt. Chem.*, 1952, **24**, 1956.
36. H. T. S. Britton, *Conductometric Analysis* (Chapman and Hall, Ltd., London, 1934), Chap. VI.
37. G. Bruni, *Z. Elektrochem.*, 1908, **14**, 701.
38. E. D. Eastman, *J. Amer. Chem. Soc.*, 1925, **47**, 335.
39. I. M. Kolthoff, *Z. anorg. Chem.*, 1920, **111**, 18.
40. *Idem, ibid.*, 9.
41. E. C. Righellato and C. W. Davies, *Trans. Faraday Soc.*, 1933, **29**, 431.
42. T. Sasaki and A. Inaba, *Bull. Chem. Soc. Japan*, 1951, **24**, 20.
43. E. Grunwald, *Analyt. Chem.*, 1956, **28**, 1112.
44. F. Gaslini and L. Z. Nahum, *Analyt. Chem.*, 1959, **31**, 989.
45. T. C. Lanzing and L. J. van der Wolk, *Rec. Trav. chim.*, 1929, **48**, 83.
46. I. M. Kolthoff, *Z. anorg. Chem.*, 1920, **112**, 156.
47. *Idem, ibid.*, 1920, **111**, 30.
48. J. Schleicher and L. Pivet, *Bol. Soc. chilena Quim.*, 1949, **1**, 9.
49. S. H. Maron, I. N. Ulevitch, and M. E. Elder, *Analyt. Chem.*, 1949, **21**, 691.
50. *Idem, ibid.*, 1952, **24**, 1068.
51. D. Ramuswamy and Y. Nayudamma, *Bull. Central Leather Research Inst.*, 1956, Madras 2, 360.
52. P. Dutoit, *Bull. Soc. chim. France*, 1910, **7**, 1.
53. I. M. Kolthoff, *Z. anorg. Chem.*, 1920, **111**, 97.
54. M. I. Lapshin, *Zavodskaya Lab.*, 1936, **5**, 1419.
55. P. Dutoit and M. Levy, *J. Chim. phys.*, 1916, **14**, 353.
56. I. M. Kolthoff, *Z. anorg. Chem.*, 1920, **112**, 156, 196.
57. W. G. Berl, *Physical Methods of Chemical Analysis* (Academic Press, Inc. New York, 1951), Vol. II, p. 80.

58. R. S. AIRS and M. P. BALFE, *Trans. Faraday Soc.*, 1943, **39**, 102.
59. A. CHESHIRE, W. B. BROWN, and N. L. HOLMS, *J. Internat. Soc. Leather Chemists*, 1941, **25**, 254.
60. N. F. HALL and W. F. SPENGEMAN, *Trans. Wisconsin Acad. Sci.*, 1937, **30**, 51.
61. T. HIGUCHI and C. R. REHM, *Analyt. Chem.*, 1955, **27**, 408.
62. M. N. DAS and D. MUKHERJEE, *Analyt. Chem.*, 1959, **31**, 233.
63. W. H. MCCURDY and J. GALT, *Analyt. Chem.*, 1958, **30**, 940.
64. J. T. PINKETON and H. T. BRISCOE, *J. Phys. Chem.*, 1942, **46**, 469.
65. N. VAN MEURS and E. A. M. F. DAHMEN, *Analyt. Chim. Acta*, 1958, **19**, 64; 1959, **21**, 443.
66. G. JANDER and K. BRODERSEN, *Z. anorg. Chem.*, 1950, **172**, 261.
67. H. T. S. BRITTON and L. W. GERMAN, *J. Chem. Soc.*, 1930, 1249, 2154.
68. E. ROTHER and G. JANDER, *Z. angew. Chem.*, 1930, **43**, 930.
69. G. JANDER and H. HEUKESHOVEN, *Z. anorg. Chem.*, 1930, **187**, 62.
70. K. PAN, S. F. LIN, and T. S. SHENG, *Bull. Chem. Soc. Japan*, **26**, 131.
71. H. T. S. BRITTON, *Conductometric Analysis* (Chapman and Hall, Ltd. London, 1934), p. 93.
72. I. M. KOLTHOFF, *Z. analyt. Chem.*, 1922, **61**, 171.
73. G. A. FREAK, *J. Chem. Soc.*, 1919, **115**, 55.
74. H. T. S. BRITTON, *Conductometric Analysis* (Chapman and Hall, Ltd. London, 1934), p. 102.
75. G. JANDER, O. PFUNDT and H. SCHORSTEIN, *Z. anorg. Chem.*, 1930, **43**, 507.
76. H. S. HARNED, *J. Amer. Chem. Soc.*, 1917, **39**, 256.
77. A. KLING and A. LASSIEUR, *Compt. rend.*, **157**, 487.
78. P. A. MEERBURG, *Chem. Weekblad*, 1918, **14**, 1054.
79. I. M. KOLTHOFF, *Z. analyt. Chem.*, 1923, **64**, 433.
80. T. KAMEDA and I. M. KOLTHOFF, *Ind. Eng. Chem. Analyt.*, 1931, **3**, 129.
81. I. M. KOLTHOFF, *Z. analyt. Chem.*, 1923, **64**, 229.
82. G. JANDER and H. IMMIG, *Z. Elektrochem.*, 1937, **43**, 207.
83. R. RIPAN and R. R. TILICI, *Z. analyt. Chem.*, 1939, **117**, 47.
84. O. PFUNDT, *Z. angew. Chem.*, 1933, **46**, 218.
85. R. R. TILICI, *Z. analyt. Chem.*, 1934, **99**, 110; 1935, **100**, 405.
86. *Idem, ibid.*, 1934, **99**, 415.
87. C. CANDIA and I. G. MURGULESCU, *Bull. Soc. chim. Romania*; *Chem. Abs.*, 1936, **30**, 7062.
88. I. M. KOLTHOFF, *Z. analyt. Chem.*, 1922, **61**, 399.
89. J. BYE, *Bull. Soc. chim. France*, 1939, **6**, 174.
90. R. R. TILICI, *Z. analyt. Chem.*, 1938, **114**, 408.
91. *Idem, ibid.*, 1935, **102**, 28.
92. P. B. MOJOIU, *Thesis*, Lausanne, 1909.
93. L. DESHUSSER and J. DESHUSSER, *Helv. Chim. Acta*, 1924, **7**, 681.
94. A. CHRETIEN and J. KRAFT, *Bull. Soc. chim. France*, (5), 1938, **5**, 1399.
95. F. H. H. VAN SUCHTELEN and A. ITANO, *J. Amer. Chem. Soc.*, 1914, **36**, 1793.
96. I. M. KOLTHOFF, *Z. analyt. Chem.*, 1923, **62**, 1.
97. *Idem, ibid.*, 161.
98. *Idem, ibid.*, 97.
99. G. HENGEVELD, *Thesis*, Lausanne, 1911.
100. H. IMMIG and G. JANDER, *Z. Elektrochem.*, 1937, **43**, 207, 214.
101. I. M. KOLTHOFF, *Z. analyt. Chem.*, 1923, **62**, 209.
102. J. H. BOULAD, *J. Soc. Chem. Ind.*, 1922, **52**, 270T.
103. J. HARMS and G. JANDER, *Z. angew. Chem.*, 1936, **49**, 106.
104. G. JANDER and O. PFUNDT, *Z. analyt. Chem.*, 1927, **71**, 417.
105. K. SHINA, K. YOSHIKAWA, T. KATO, and Y. NOMIZO, *J. Chem. Soc. Japan, Pure Chem. Sect.*, 1954, **75**, 46.
106. V. V. UDOVENKO and G. B. PASOVSKAYA, *Zhur. analit. Khim*, 1952, **7**, 158.
107. G. JANDER and K. F. JAHR, *Z. angew. Chem.*, 1931, **44**, 977.

108. P. Dutoit and Reeb, *Chem. Ztg.*, **37**, 469; *Chem. Abs.*, 1914, **8**, 472.
109. H. Fehn, G. Jander and O. Pfundt, *Z. angew. Chem.*, 1929, **42**, 158.
110. I. M. Kolthoff, *Z. analyt. Chem.*, 1923, **62**, 332.
111. R. R. Tilici, *Z. analyt. Chem.*, 1936, **107**, 111.
112. S. Demjanovski, *J. Soc. Phys. Chim. Russe Univ. Leningrad*, 1924, **55**, 327.
113. R. A. Robinson and H. T. S. Britton, *J. Chem. Soc.*, 1931, 2817.
114. P. Raff and W. Brotz, *Z. analyt. Chem.*, 1951, **133**, 241.
115. W. W. Wendlandt, *Chemist-Analyst*, 1957, **46**, 8.
116. G. A. Schöberl, *Z. analyt. Chem.*, 1948, **128**, 210.
117. W. Scheele and C. Gensch, *Kautschuk u. Gummi*, 1955, **8**, WT 55.
118. *Idem, ibid.*, 1954, **7**, WT 122.
119. J. C. Bailar, *The Chemistry of Co-ordination Compounds* (Reinhold Publishing Corp., New York), 1956, pp. 113–118.
120. P. Job, *Compt. rend.*, 1920, **170**, 731.
121. A. K. Dey, *Nature*, 1946, **158**, 95.
122. J. Kamecki and J. Wolny, *Roczniki Chem.*, 1948, **22**, 48.
123. B. C. Halder, *J. Indian Chem. Soc.*, 1947, **24**, 503.
124. H. T. S. Britton and W. E. Battrick, *J. Chem. Soc.*, 1932, 196; 1933, 5.
125. S. G. Shuttleworth, *J. Soc. Leather Trades' Chemists*, 1954, **38**, 110, 232.
126. *Idem, ibid.*, 1950, **34**, 186.
127. *Idem, J. Amer. Leather Chemists' Assoc.*, 1950, **45**, 447.
128. *Idem, Leather Ind. Research Bull.*, (S.A.), 1943, **2**, Circ. No. 24, 357.
129. *Idem, J. Internat. Soc. Leather Chemists.*, 1946, **30**, 342.
130. E. R. Theis and E. J. Serfass, *J. Amer. Leather Chemists' Assoc.*, 1934, **29**, 543.
131. Y. Ono and Y. Inoue, *Reports Himeji Inst. Technol.*, 1955, **5**, 53.
132. H. T. S. Britton and E. N. Dodd, *J. Chem. Soc.*, 1932, 1951.
133. R. Gaugin, *Analyt. Chim. Acta*, 1949, **3**, 489.
134. J. L. Hall, J. A. Gibson, P. R. Wilkinson and H. O. Phillips, *Analyt. Chem.*, 1954, **26**, 1484.
135. F. Vydra and M. Karlik, *Chem. Listy*, 1956, **50**, 1749, 1754.
136. A. E. Martell, *J. Chem. Educ.*, 1952, **20**, 270.
137. J. Harms and G. Jander, *Z. Elektrochem.*, 1936, **42**, 315.
138. H. Kubota and J. Surak, *Analyt. Chem.*, 1959, **31**, 283.

Chapter V

High Frequency Conductometric (Impedimetric) Titrations

T. S. BURKHALTER

1. Introduction

The measurement of the conductivity of electrolytic solutions has been a well established analytical tool since the early 1920's, and is usually made by imposing a potential difference on two platinum electrodes immersed in the solution, the resistance of which is then determined by making it one arm of a bridge arrangement. The serious, and inherent, disadvantage of this method is that electrodes in direct contact with a solution suffer polarisation and fouling, which may introduce spurious resistances that cannot be interpreted in terms of the composition of the solution. Zahn[1], who was investigating conductometric dispersion in the light of the frequency theories of Debye and Falkenhagen[2], was the first to make conductance measurements on solutions of electrolytes by means of electrodes that were not in contact with the solution: he measured the transformer energy loss at a frequency of 300 Mc/sec when a beaker containing an electrolytic solution was placed inside the core of the primary, but he apparently did not realize the analytical possibilities of his work. In 1938, Blake[3] discussed the possibility of making conductance measurements at high frequencies with electrodes outside the cell, but there was no general acceptance of this idea until 1946, when both Jensen[4] and Blake[3] independently published conductometric titration curves made with electrodes that had no physical contact with the solution being titrated. The uniqueness of this technique lies in the fact that a chemical system absorbs electromagnetic energy of radio-frequency through the walls of the containing vessel and stores it. The resulting energy transformation is reflected in the operation of the generator that produces the electromagnetic field and, if some electrical parameter of the generator is measured, it is found to be a function of the magnitude of the energy absorption and hence of the composition of the chemical system. The advantage of this type of measurement over conventional conductometric measurements is that, as the electromagnetic energy is transmitted through the

References p. 254

walls of the container, there is no physical contact between the chemical system and the electrodes, and electrode polarisation and fouling are avoided. Further, the electrical parameter used to indicate the energy change (the sensible indicator) can be measured with a direct-reading meter and the tedious operation of bridge balancing is eliminated.

2. Definition of Terms

The term conductance has been so universally misused in connection with this analytical technique that a clarification of terms is necessary before any explanation of the principles of the method can be attempted. *Conduction* involves the movement of electric charges under the influence of a potential gradient and against a resistance. *Conductance* is the reciprocal of ohmic resistance and is always associated with an energy loss. *Reactance* involves the temporary displacement of electric charges as an electric field is built up and relaxed; there is no energy loss with a pure reactance. The vector sum of resistance and reactance is known as *impedance, i.e.*,

$$Z \text{ (impedance)} = \sqrt{R^2 \text{ (resistance)} + X^2 \text{ (reactance)}}$$

When electromagnetic energy is transmitted through a cell and chemical system of the type used in so-called "high-frequency conductometric" methods, both a resistive and a reactive component are always present. Although the generator circuit and cell geometry may be so designed that the electrical parameter used as the sensible indicator is primarily a function of either the resistive or the reactive component, no practical instrument yet described measures only the resistance, or the reactance, of the chemical system. In every case, the measured parameter is a rather complex function of the total impedance of the cell and chemical system. For this reason, the term "high-frequency conductance", when used to describe the property measured, is a most unfortunate misnomer and, because of its widespread usage, has introduced considerable and needless confusion. Throughout this chapter the perfectly valid generic term of *impedance* will be used instead of "high-frequency conductance", and the term *impedimeter* will be used to designate all analytical instruments for making measurements of the conductance type by means of electrodes placed outside the cell. It should also be clearly understood that the measured value is not a linear function of the impedance *per se*, but may reflect a disproportionate amount of either the resistive or the reactive component of the impedance. The fact that the measured value does not indicate the magnitude of any single property of the chemical system does not destroy its utility as an analytical tool. By means of suitable circuits, the measured value does become a nearly linear function of the composition of the chemical system (over a relatively narrow concentration range), and is used advantageously as the indicator in titrimetry; it may also be used analytically as though it were a measure of some purely fundamental property of the chemical system. The term *cell* is defined as the container for the chemical system, together with any metallic plates or bands serving as electrodes and the leads needed to make electrical connection to the field generator circuit. The term *cell system* is used to designate the cell filled with a chemical system. The terms *conductance, specific conductance, resistance, reactance, impedance, susceptance* and *admittance* are used in their usual senses as defined in the nomenclature of electrochemistry and electronics. The term *high-frequency conductance* is occasionally used to indicate that the conductance measurement was made at frequencies above 1 Mc/sec, and *low-frequency conductance* means a conductance measurement at the more conventional frequency of one thousand cycles per second.

3. Theory and Principles

(a) Fundamental Theory

The essential equipment for impedimetric methods of analysis is an electromagnetic field generator, usually capable of working between 1 and 300 Mc/sec and equipped with a metering device for measuring some electrical parameter of the field generator, and a cell that can be coupled to the field. The generator usually consists of a simple low-power vacuum-tube (valve) oscillator. The chemical system to be analysed is brought into the field of the oscillator by placing the cell inside the core of one of the coils in the circuit, or by mechanical connection of the cell electrodes with the oscillator circuit. In either position, the transmission of the field through the cell system reflects both a resistive and a reactive component of the cell system impedance back into the oscillator; both components effect the operation of the oscillator. The resistive component results primarily in a power loss in the oscillator, and a change in the magnitude of the plate, grid, and cathode currents as well as, in certain cases, a change in the field frequency. The reactive component results principally in a change of the natural resonant frequency of the field generator and, to a lesser extent, in a change of the magnitude of the currents in the oscillator circuit.

As the composition of the chemical system is changed, the magnitude of the impedance reflected into the oscillator is changed, and this change may be followed by measuring the value of any of the electrical parameters mentioned above. The course of a titration may be followed in this way: a plot of the characteristic measured *vs.* volume of titrant added results in a curve with a break at the stoichiometric end point, resembling in most instances the type of curve obtained in a conventional conductometric titration. Numerous attempts have been made to equate the magnitude of the measured parameter to the conductance of the chemical system; although this has been successfully done by a number of investigators[5, 6, 7] on a purely empirical basis for a given cell and instrument, no general equation relating the conductance or reactance of the chemical system to any measured parameter of the field generator has yet been derived. The magnitude of both the resistive and the reactive components of the impedance of the cell system can, however, be determined with any of a number of commercially available high-frequency impedance-measuring instruments. From these values and by the use of electrically equivalent circuit theory, a fairly good approximation of the actual value of the resistance and the reactance of the chemical system can be calculated[8]. The way in which the cell system impedance changes with a change in the impedance of the chemical system allows quite accurate prediction of the nature of the response of any given instrument.

(b) Cell System Impedance

Although a number of designs for impedimetric cells have been described in the literature, the one that lends itself most readily to equivalent circuit

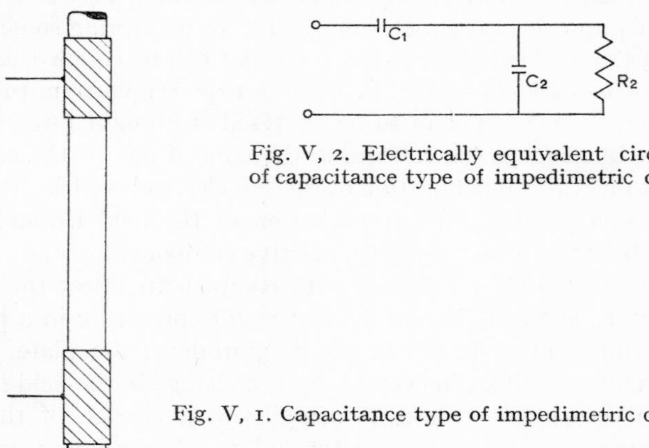

Fig. V, 2. Electrically equivalent circuit of capacitance type of impedimetric cell.

Fig. V, 1. Capacitance type of impedimetric cell.

analysis is that of a glass tube round which two metallic bands with leads for electrical connection (Fig. V, 1) have been placed. An approximate electrical equivalent of this would be a condenser in series with a parallel branch consisting of a resistor and a second condenser, Fig. V,2. C_1 represents a condenser consisting of the metallic band serving as one plate of a parallel plate condenser; the glass wall of the tube serves as the dielectric material between the plates, and the conducting solution within the tube serves as the second plate of the condenser. Actually, there are two such condensers in the cell system but, for the purpose of calculation, they may be considered together as C_1. C_2 and R_2 represent the capacity and resistance of the chemical system. It is the impedance of this total circuit that is reflected into the field generator. The nature of the variations in the components of this total impedance may be examined qualitatively by considering the extremes of the possible changes of the properties of the chemical system.

The resistive component of the impedance of the circuit is considered first: if R_2 becomes zero, there is no resistive component in the circuit, and the oscillator is dealing with a purely reactive impedance (no power loss). If R_2 becomes very large, *i.e.*, approaches infinity, the resistance is effectively short-circuited by C_2; again, the resistive component is effectively zero,

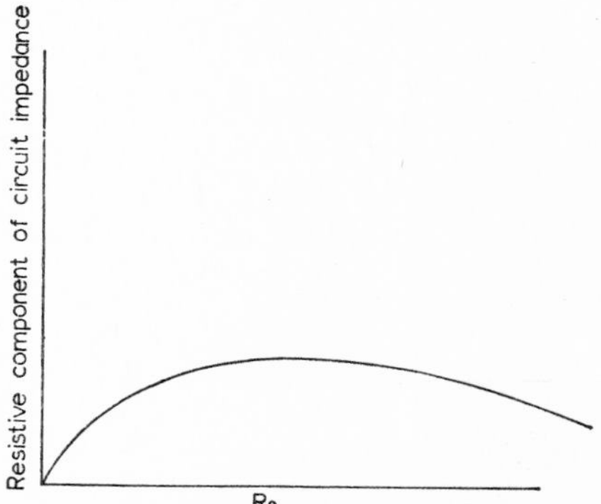

Fig. V, 3. General shape of response curve of a resistance type of impedimeter.

and the oscillator is dealing only with a reactive impedance. At all values of R_2 between zero and infinity, there exist both reactive and resistive components, which result in a power loss in the circuit. It follows then that, as R_2 varies from zero to infinity, the resistive component of the circuit impedance must vary from zero, through a maximum, and return again to zero. The response curve of an impedimeter that measures principally the resistive component will be of the shape shown in Fig. V, 3, with the maximum of the curve depending upon the conductance of the chemical system.

The nature of the relationship between the reactance of the cell system and the conductance of the chemical system may also be elucidated qualitatively by examining the possible extremes of R_2: when R_2 is zero, the impedance of the circuit is purely reactive and the reactance of the cell system is a minimum, because C_2 is effectively short-circuited and the impedance

References p. 254

of the circuit is that of C_1. If R_2 becomes infinitely large, all the current in the parallel branch must pass through C_2 and, again, there exists only a reactive impedance, but the value of this reactance will be the maximum, because the current passes through C_1 and C_2 in series. The response curve of an impedimeter that measures principally the reactive component must be asymptotic in form, with the maximum response due to the difference in value of C_1 alone and C_1 and C_2 in series (Fig. V, 4).

This simple picture is somewhat complicated by the fact that, for any given chemical system, the reactance and resistance of the system are interrelated; both, however, change in the same direction with a given change in the composition of the system, and the general shape of the response curve is not altered by this interdependence. The picture is still further complicated by the dependence of C_1 upon the composition of the chemical system; in the argument just used, C_1 was considered constant and independent of the composition of the chemical system, but this assumption is not valid for the following reason: the second plate of C_1 is the chemical system, which is usually a solution of electrolytes, and the value of C_1 is dependent upon the area of this plate; the plate must be a conductor, and its effective area must therefore be a function of the concentration of the conducting particles in the solution, that is, of the ionic concentration. This dependency of the value of C_1 upon the composition of the system is extremely complex, and no electrical equivalent has been suggested.

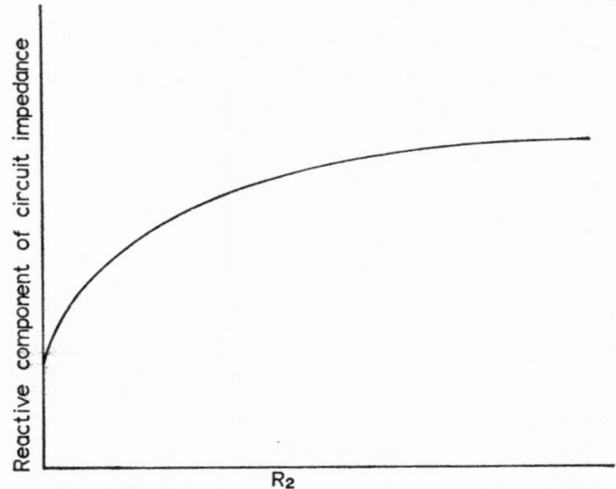

Fig. V, 4. General shape of response curve of a resistance type of impedimeter.

(i) Resistive Component

If the assumption of the constancy of C_1 is accepted, it is possible to derive equations relating both conductance and reactance of the cell system to the low-frequency conductance of the chemical system. The following argument is a composite of those of Hall[8] and Reilley[9]: the equivalent circuit of Fig. V, 2 may be simplified still further to a single parallel branch (Fig. V, 5), where R_p and C_p represent the net parallel impedance of the cell system; this parallel equivalent circuit is derived from Fig. V, 2 by the standard

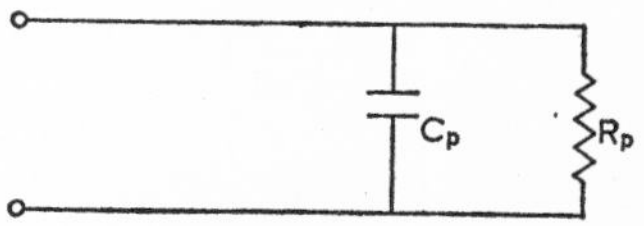

Fig. V, 5. Parallel electrically equivalent circuit of capacitance type of impedimetric cell.

method of summation of admittances and rationalising into real and imaginary terms.

$$Y_p = G_p + JB_p, \tag{1}$$

$$Y_p = \frac{(1/R)W^2C_1^2}{1/R^2 + W^2(C_2 + C_1)^2} + J\,\frac{WC_1/R^2 + W^3C_1C_2^2 + W^3C_1^2C_2}{1/R^2 + W^2(C_1 + C_2)^2}, \tag{2}$$

where

Y_p = admittance of the circuit,
G_p = conductance term $(1/R)$, the real part of admittance,
J = operator $(-1)^{1/2}$,
B_p = susceptance term (equal to WC_p), the imaginary part of admittance,
W = $2\pi f$, where f is the frequency,
R = resistance offered by the chemical system (equal to $1/k$, where k is the low-frequency conductance of the system),
C_1 = capacitance due to chemical system.

Evaluating the conductance in terms of k, the low-frequency conductance of the chemical system,

$$G_p = 1/R_p = \frac{kW^2C_1^2}{k^2 + W^2(C_2 + C_1)^2}. \tag{3}$$

From equation (3), as C_1 increases, the value of G_p increases, and as C_1 approaches infinite capacitance, the high-frequency conductance of the cell system approaches, as a limit, the low-frequency conductance of the chemi-

References p. 254

cal system. Further, it is seen that G_p is a unique function of k, the low-frequency conductance. Thus, the high-frequency conductance of the cell system is independent of the nature of the chemical system, and is dependent only on the low-frequency conductance of the system. Thus, a plot of G_p *vs.* K (specific conductance) for any series of solutions of electrolytes with the same solvent and the same cell at a fixed frequency should fall on a single curve, similar in shape to that of Fig. V, 3; that such is the case has been experimentally verified.

The position of the peak is important from the standpoint of the shape of titration curves. Differentiation of equation (3) for maxima leads to

$$k_{\text{peak}} = W(C_2 + C_1) \tag{4}$$

Equation (4) shows that the conductivity of the chemical system corresponding to the peak of the G_p curve increases directly with frequency; thus, a

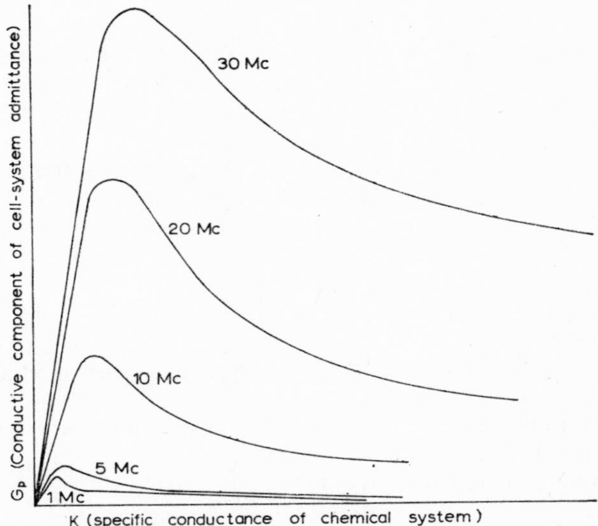

Fig. V, 6. Family of response curves for resistance type of impedimeter at various frequencies.

family of curves exists for any given cell system with a common solvent, the peak of the curve appearing at higher specific conductances with increasing frequency. Further, substitution of equation (4) into equation (3) gives the relationship

$$G_{p(\text{peak})} = \frac{WC_1^2}{2(C_1 + C_2)} \, . \tag{5}$$

Thus, the magnitude of $G_{p(\mathrm{peak})}$ increases with increasing frequency. Fig. V, 6 illustrates both of these trends.

The effect of different solvents upon the variation of the conductance G_p of the cell system is also shown by equation (4). The value of C_2 decreases as the dielectric constant of the solvent is lowered, hence the conductance of the solution will be less at $G_{p(\mathrm{peak})}$ for solutions with solvents of low dielectric constants. This again leads to a family of curves at fixed frequencies for diverse solvent systems (Fig. V, 7).

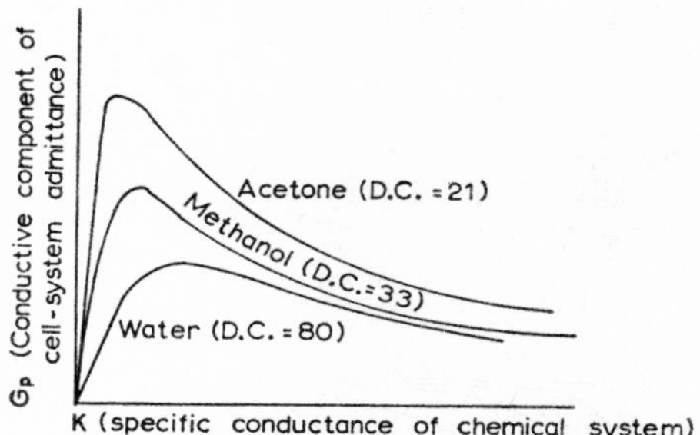

Fig. V, 7. Family of response curves for resistance type of impedimeter with various solvents.

(ii) Reactance Term

Evaluation of the imaginary term of the admittance equations (1) and (2) as a function of the low-frequency conductance of the system leads to

$$C_p = \frac{C_1k^2 + W^2C_1C_2^2 + W^2C_2C_1^2}{k^2 + W^2(C_1 + C_2)^2}. \qquad (6)$$

A number of important conclusions and relationships can be drawn from equation (6): first, the reactance of the cell system is a unique function of the conductance of the system; secondly, the value of the capacitance of the cell system varies from a maximum of C_1, as the conductance of the system becomes infinitely large, to a minimum of $C_1C_2/(C_1+C_2)$, as the conductance of the system approaches zero. This inference has already been established in the qualitative approach described earlier. The difference in the extreme values of capacitance (ΔC_p) is given by

$$\Delta C_p = C_1^2/(C_1 + C_2). \qquad 7)$$

Thus, in instruments built around reactance measurements, C_1 should be as large as is practicable and C_2 as small as practicable, in order to provide for large changes in capacitance. A still more important relationship may be

References p. 254

derived by solving equation (6) for the value of k at the mid-point of the curve, where the capacitance is half way between its extreme values.

$$C_1 - \tfrac{1}{2}\Delta C_p = \frac{C_1k^2 + W^2C_1C_2^2 + W^2C_2C_1^2}{k^2 + W^2(C_1 + C_2)^2}, \tag{8}$$

and

$$k_{(\text{mid-point})} = W(C_1 + C_2). \tag{9}$$

The peak high-frequency conductance and the mid-point on the capacitance curve occur at the same low-frequency conductance point, because equation (9) is identical with equation (4); further, the slope of the curve near the mid-point is found to be inversely proportional to the frequency. Consideration of these relationships leads again to a single reactance curve at constant frequency, but to a family of curves for a given cell with the same solvent at various frequencies (Fig. V, 8).

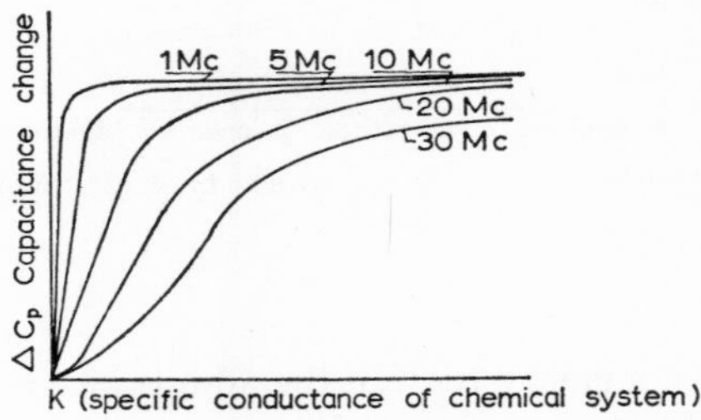

Fig. V, 8. Family of response curves for resistance type of impedimeter at various frequencies.

(*iii*) *Coil Type of Cell*

A similar theoretical treatment for the case of a cell placed inside the core of a coil in the oscillator circuit has been given by Haskins, Heller, and Miller[10] but, as the treatment is much more complex than that for the condenser type of cell and leads essentially to the same conclusions, it is not included here.

(*c*) *Principles of Analysis*

In any instrumental analytical approach, two criteria, sensitivity of response and stability (or reproducibility) of measurement, are particularly important. It is almost universally true that, as the limit of sensitivity is approached, stability of measurement decreases; it is equally true that, at the maximum sensitivity consistent with acceptable stability, the versatility, or range of applicability to diverse systems, is least, although the impedimetric method of analysis is not unique in this respect. It becomes necessary, therefore, to understand the factors that affect these important criteria. Two operative variables, frequency and cell design, that affect sensitivity, stability and versatility will be considered here, while a third, oscillator design, will be considered under the topic of apparatus.

(d) Types of Impedimeters

(i) Resistance Type of Impedimeter

Inspection of Fig. V, 6 and equations (3) and (4) reveals relationships pertinent to the resistance type of impedimeter. If the effect of frequency is considered first, it will be noted from Fig. V, 6 that the value of G_p is greater at higher frequencies for any given value of k; this means that the increment of instrument response for a given increment of system composition change (sensitivity) will be greater at the higher frequencies. It should also be noted that the value of k that corresponds to $G_{p(\mathrm{peak})}$ increases with increasing frequency; this shows that the more sensitive portion of the curve (before the maximum) covers a wider range of conductivity of the system, and permits sensitive measurements at greater concentration with increasing frequency. On the other hand, it is an inescapable fact that the stability of an electronic oscillator decreases at higher frequencies, with the result that the measurement of an electrical parameter becomes increasingly difficult; the best frequency for satisfactory operation will thus be at some point between maximum stability and maximum sensitivity, and will depend upon the parameter to be measured.

Cell geometry also plays an important part in the sensitivity and useful range of operation and, to a lesser extent, in the stability of the system. It can be seen from equation (5) that $G_{p(\mathrm{peak})}$ increases directly as the square of C_1 and inversely with the sum of C_1 and C_2. For increased sensitivity, C_1 should be made as large as possible and C_2 small. This may be effected by the use of wide metallic bands, relatively for apart, surrounding a tube of small bore and thin walls. Conversely, it can be seen that $k_{(\mathrm{peak})}$, which determines the useful concentration range, varies directly with both C_1 and C_2. The value of C_2 may be conveniently increased by the use of tubing of wide bore, and by making the distance between the two bands as small as is practicable. For maximum sensitivity, or for a wide range of concentration, C_1 should be large, but the sensitivity is increased by decreasing the value of C_2, only at the expense of limiting the concentration range.

(ii) Reactance Type of Impedimeter

Fig. V, 8 and equations (6) to (9) show the effect of frequency and cell geometry upon the sensitivity and versatility of the reactance type of impedimeter. The effect of higher frequencies is to increase the useful concentration range, but to decrease the sensitivity. It may be noted that the one-megacycle curve has an extremely steep slope, but that the concentration range covered by the sensitive portion is quite narrow; on the other hand, the slope of the thirty-megacycle curve is much flatter, but covers an appreciably greater concentration range. As regards cell geometry, equation (7) shows that the difference in the extreme values of capacitance, and thus the ultimate sensitivity, is directly proportional to C_1^2 and inversely proportional to the sum of C_1 and C_2; maximum sensitivity is attained then when C_1 is large com-

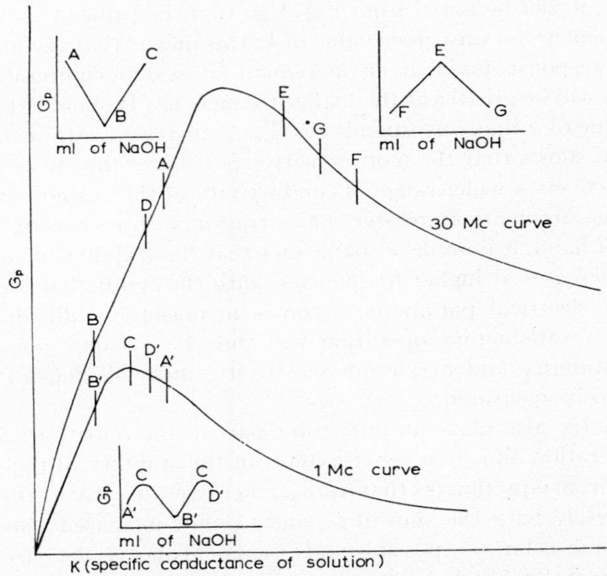

Fig. V, 9. Titration curves of HCl with NaOH at one and thirty megacycles per second.

pared with C_2. Equation (9) shows that the sensitive range of concentration increases in proportion to the sum of C_1 and C_2. Here again, large values for C_1 improve both sensitivity and range, while an increase in the value of C_2 gives an increase in range at the expense of sensitivity, and *vice versa*.

(e) *Shape of Titration Curves*

The effect of frequency and of concentration of the solution upon the shape of titration curves may be explained on the basis of Fig. V, 9, which is a partial reproduction of Fig. V, 6. Several investigators[11, 12, 13] have reported that the normal V-shaped curve for the titration of a strong base with a strong acid is inverted or distorted to a winged V (see Fig. V, 9) as the frequency or initial concentration is increased. It becomes apparent that this should be expected, if the course of a titration of hydrochloric acid with sodium hydroxide at one and at thirty megacycles is followed. The concentration of the initial acid solution is such that the G_p for the cell system at the beginning of the titration lies to the left of the $G_{p(\mathrm{peak})}$ on the thirty-megacycle curve: (points A and A.') As sodium hydroxide is added until the equivalence point (B and B′) is reached, the conductance of the solution steadily decreases and, at 30 Mc/sec, the high-frequency conductance of the cell system will also decrease, but at 1 Mc/sec the G_p value will at first increase to point C, and then decrease to point B′. After the equivalence point, addition of excess of sodium hydroxide will cause a rise in the conductance of the solution and a corresponding rise in the G_p value on the thirty-megacycle curve to point D; on the one-megacycle curve, however, the rise in conductance will cause a corresponding increase in the G_p value only until the $G_{p(\mathrm{peak})}$ (point C) is reached and further addition of sodium hydroxide beyond that point will result in a decrease of G_p to point D′. The resulting titrat on curves, shown in the insets, will be V-shaped for the 30-megacycle titration and shaped like a winged V for the one-megacycle titration.

If a hydrochloric acid solution is again titrated with sodium hydroxide at thirty megacycles, but this time with the initial concentration such that the G_p value lies to the right of $G_{p(\mathrm{peak})}$ on the thirty-megacycle curve at point F, an inverted V-shaped titration curve will be obtained. Starting at point F, as sodium hydroxide is added the conductance of the solution will steadily decrease, while the G_p value of the cell system will steadily increase to the stoichiometric end point at E. Addition of sodium hydroxide beyond the end point will result in an increase in the conductance of the solution but a decrease in the G_p value for the cell system to point G, with the resulting inverted V-shaped titration curve shown in the inset.

All titration curves from a reactance type of impedimeter will be V-shaped, because the capacity of the cell system increases with increase in conductance and exhibits no maximum; the slope of the sides of the V will, however, show considerable curvature if the titration is carried out in the region of maximum curvature (Fig. V, 8); or the slopes will be very flat if the titration is conducted at a point beyond the region of maximum curvature. The preparation of a family of curves relating instrument response to K at various frequencies for any given combination of impedimeter and cell system allows the prediction of the shape of any titration curve in the same solvent.

References p. 254

4. Apparatus

The essential apparatus for impedimetric measurements is simple, and can be assembled relatively cheaply from components usually available in the laboratory. A radio-frequency oscillator with a metering system and a device for coupling the oscillator field to the sample system are needed for the circuit. It must be emphasized, however, that the design, lay-out, and construction of a stable electronic oscillator with loading characteristics suitable for impedimetric measurements is an operation that requires a considerable knowledge of electronics and some experience in the art of assembling electronic circuits. The reader is warned not to expect satisfactory performance from an impedimeter that has been constructed without careful attention to fundamental radio-frequency techniques. Although the total number of possible combinations of oscillator circuits, metering circuits and cell connections is so large that a comprehensive discussion of this subject must be considered beyond the scope of this chapter, a brief discussion of fundamentals is given below.

(a) Oscillator Circuits

Several factors must be considered in the choice of an oscillator circuit: simplicity in design with the minimum of wiring is desirable; the oscillator

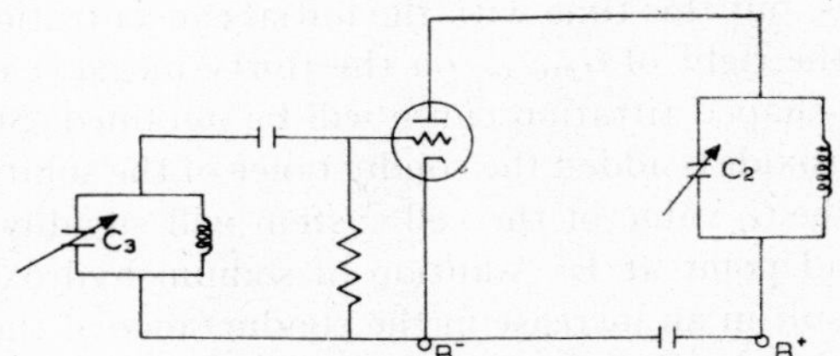

Fig. V, 10. Tuned-plate tuned-grid oscillator circuit.

must be stable in operation and yet be loaded readily to produce good sensitivity, and the loading characteristics should be nearly linear over a wide range of load; as oscillators of high sensitivity can be loaded out of oscillation quite readily, another requirement is that oscillatory conditions should be easily restored. Two basic types of oscillators have been shown to be most useful for impedimetric measurements.

The tuned-plate tuned-grid oscillator[14] shown schematically in Fig. V, 10 is extremely sensitive to grid or to plate loading. It has two tuned circuits, which must be adjusted to resonate at approximately the same frequency for oscillation to take place. Under steady conditions of light load, the frequency stability of the oscillator is good, but de-tuning of either tank circuit by changing the load (as must occur during a titration) appreciably decreases the stability of the frequency. Under light load, the loading characteristic of the oscillator is very sensitive and linear but, with medium load changes, the oscillator goes out of oscillation because of the de-tuning effect of the circuit being loaded. Oscillation can be restored quite readily by re-tuning either tank circuit with the variable condensers, C_2 or C_3. The oscillator must be very carefully shielded from external influences such as body capacity, because feed-back is attained through the inter-electrode capacitance of the vacuum tube. The tuned-plate tuned-grid oscillator has proved adaptable to chemical analysis because of its high sensitivity, the relative ease with which it can be brought back into a state of oscillation, and because, near the resonance point, its loading curve has a steep portion in which the response is virtually linear.

The Colpitts-Clapp oscillator[15], shown in Fig. V, 11, has the greatest frequency stability of any vacuum-tube oscillator, except the crystal-controlled type. This is achieved through a single low impedance series-tuned tank (C_1 and L_1) grid circuit. Spurious thermal excitations within the tube are swamped out by the large condensers C_2 and C_3, which are connected in parallel with the inter-electrode capacitance of the tube. Resistive load

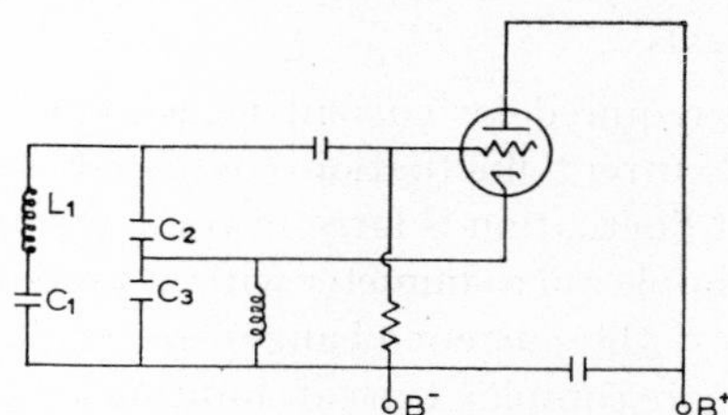

Fig. V, 11. Colpitts-Clapp oscillator circuit.

characteristics of this oscillator are inferior to those of the tuned-plate tuned-grid circuit, because the frequency stability of the oscillator is based upon its maintaining a low power-loss factor; the capacitive load ability of the oscillator is, however, exceptionally good, and the capacitive loading curve is nearly linear and very sensitive over a wide range. The oscillator is exceedingly simple to construct and requires the minimum of shielding. Because of its excellent capacitive loading characteristics, the Colpitts-Clapp oscillator is recommended for use with the reactance type of impedimeter.

Crystal-controlled oscillators depend upon the piezoelectric effect of quartz, and the stability of their frequency is decidedly superior to that of

References p. 254

other types. A quartz crystal may be substituted for the grid tank in the tuned-plate tuned-grid or Colpitts circuits with a resulting improvement in stability and, for certain types of metering circuits to be described later, the crystal-controlled oscillator can be used most advantageously.

The desirability of making impedimetric measurements at the higher frequencies has been previously discussed (p. 225) and, although either the Colpitts or the tuned-plate tuned-grid oscillator can be operated at frequencies as high as one hundred megacycles, the stability attainable at this frequency makes them unsatisfactory for impedimetric measurements. A quarter-wave-length concentric line oscillator operating at three hundred and fifty megacycles has been described by Blaedel[16]; although he obtained satisfactory results, the concentric line oscillator is not commonly used.

(b) *Metering Circuits*

The type of metering circuit required depends to a large extent upon the electrical parameter chosen as the indicator, and may vary in complexity from a simple microammeter to a potentiometric recorder fed by a frequency discriminator. The plate current, or the oscillator frequency, is usually chosen as the characteristic to be measured as a function of the composition of the system, although grid current, grid voltage, capacitance re-tune, voltage across the load and other values have been used successfully.

(*i*) *Current Measurements*

The type of circuit required for current measurement depends upon the relative value of the current fluctuation compared with the steady-state current: if the current fluctuation is large in comparison with the current in the steady state, a simple microammeter with a standard shunting arrangement is satisfactory; if the current changes are small compared with the steady-state value, a more complex type of shunting arrangement is required. The so-called "voltage bucking" or "zero shunt" circuit shown in Fig. V, 12 works as follows: in the steady state, the potential drop across R_2 is balanced by an equal and opposite potential drop across R_3 and part of R_4 by adjusting the contact point of R_4 until no current flows through the meter M, which is a zero-centered d.c. microammeter with a range that will cover all current changes occurring in the course of a single titration; when the shunt circuit is properly adjusted, any increase or decrease in the plate current will be registered on the meter. C_2 and C_3 are capacitive shunts to prevent radio-frequency voltages from entering the metering circuit, and R_6 is a protective shunt to prevent damage to the meter whilst R_4 is being adjusted.

(ii) *Voltage Measurements*

Grid bias voltage has been used by Hall[17] as the titrimetic indicator for a very simple and most practical impedimeter; a standard commercial valve voltmeter of suitable range is adequate for such measurements. Jensen *et al.*[18] and Beaver *et al.*[19] have described methods involving the measurement of radio-frequency voltage across tuned circuits containing the cell system; but such measurements require much more elaborate circuits, and can be justified only for the particular purpose for which they were designed.

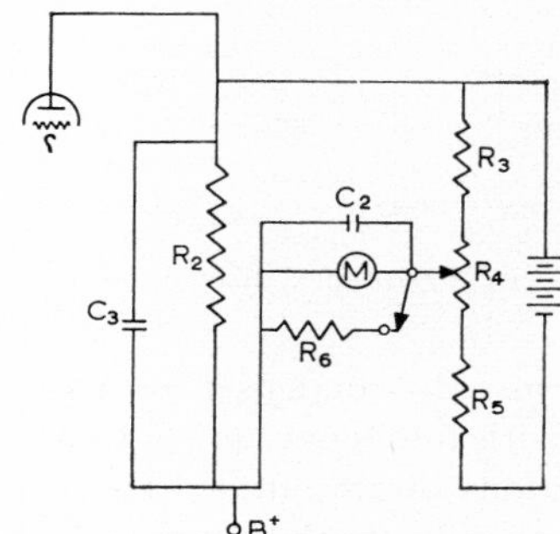

Fig. V, 12. "Voltage bucking" metering circuit.

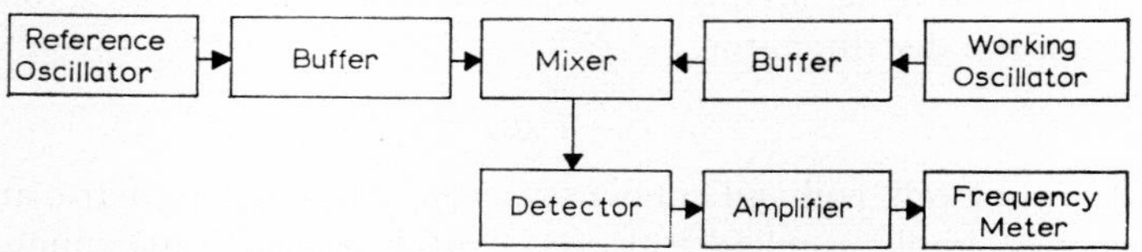

Fig. V, 13. Block diagram of heterodyne system for measuring frequency change.

(iii) *Frequency Measurements*

Direct-reading commercial radio-frequency meters are seldom accurate enough to determine small frequency changes. In the field of impedimetry, frequency changes are usually measured by the heterodyne method, or by a frequency-discrimination circuit. In the heterodyne method, two frequencies are fed into a mixer tube, and the difference between the two is measured; West[20] has described an instrument in which this principle is used: the West impedimeter, shown in Fig. V, 13, consists of two oscillators (a "working" oscillator containing the cell system, and a "reference" oscillator), a mixer stage, a detector, and an audio-frequency meter. The reference oscillator operates at a fixed frequency that is only slightly different from the steady-state frequency of the working oscillator. The detector stage

References p. 254

isolates a frequency that is equal to the difference between the frequencies of the working and reference oscillators, and is subsequently measured by the audio-frequency meter. When a change in the composition of the system causes a change in the frequency of the working oscillator, there is a corresponding change in the difference frequency, which is accurately indicated by the frequency meter.

The simplest type of discriminator is a resonant circuit tuned to the frequency of the oscillator (Fig. V, 14). Over a limited range of frequency,

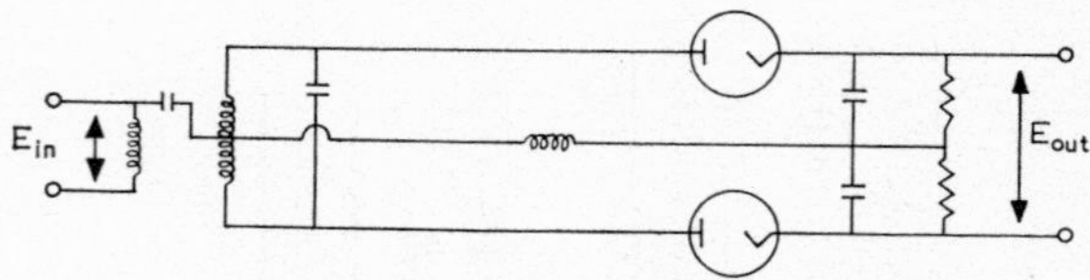

Fig. V, 14. Frequency discriminator circuit.

the voltage produced across the condenser, or across the inductance, varies approximately linearly with frequency, so that variations of frequency produce proportional variations of amplitude. The voltage output of the discriminator may be used to indicate the magnitude of the frequency change; alternatively, it is possible to measure the capacitance needed to re-tune the oscillator to its original frequency, which will be indicated by zero voltage output from the discriminator.

(c) Cells

Two types of cell, coil and condenser, are used in impedimetric analysis. The former is simply a cylindrical glass container, which fits snugly inside the core of one of the coils in the oscillator circuit; coupling to the field occurs through the capacitance that exists between individual turns on the coil; as this coupling is usually very loose, the oscillator loading is light. There is no choice in the shape of these cells because, of necessity, they must fit smoothly against the inside walls of the coil core. Condenser cells have been previously described in the section on principles (Fig. V, 1) and consist basically of a glass cylinder, which serves as the container for the chemical system, and two metal bands attached firmly to the outside of the cylinder. The cell system is coupled into the oscillator circuit by means of leads connected to the metal bands.

(d) Classification of Practical Impedimeters

The various combinations of the basic units can be divided into three categories, typical examples of which are now given.

(*i*) *Type 1. Q-Metric Impedimeters*

This type includes all impedimeters in which the cell system does not form part of the frequency-controlling elements of the oscillator circuit. Blake[21], Beaver[19], Hall[17], and Jensen[18] have described impedimeters of this class. Blake's "Rectified Radio-frequency Conductometric Titrator" (Fig. V, 15), which is an excellent example of this type, consists of a Hartley oscillator with the cell capacitively coupled as the plate load of the oscillator. The radio-frequency current passing through the cell is rectified and measured

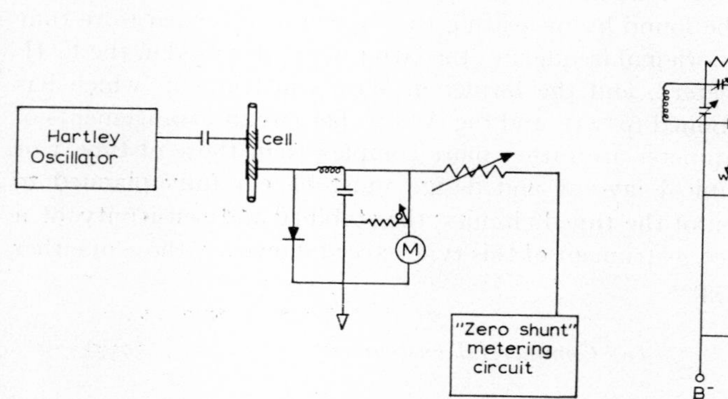

Fig. V, 15. Blake's rectified radio-frequency conductometer.

Fig. V, 16. Jensen's (Type 2) impedimeter.

as the sensible indicator. The principal advantage of this type of instrument is that it is easy to assemble. Almost any commercial radio-frequency oscillator can serve as the field generator; rectification is effected by crystal diodes, and a precision microammeter is suitable for measuring the rectified current.

(*ii*) *Type 2. Resistance Impedimeters*

Type 2 impedimeters include all those instruments in which the cell system constitutes part of the frequency-controlling elements of the oscillator, and in which the electrical parameter to be measured depends mainly on the resistive load on the oscillator; the plate and grid currents of the oscillator are the parameters most usually measured. This has been the most popular type of instrument, and a number of modifications have been described by Jensen[4], Arditti[22], Anderson[11], Milner[23], and others. Jensen used a tuned-plate tuned-grid oscillator and measured the plate current as the load indicator (Fig. V, 16); his cell was a "Pyrex" test-tube seated firmly inside

References p. 254

the core of the coil of the plate tank circuit. The instrument is extremely sensitive to small changes in the composition of dilute solutions, but suffers from the disadvantage that, unless elaborate shielding precautions are taken, it is equally sensitive to the body capacity of the operator.

(iii) *Type 3. Reactance Impedimeters*

Type 3 comprises impedimeters in which the cell system forms part of the frequency-controlling elements of the oscillator and the shift of frequency is used as the sensible indicator. The change in frequency may be measured directly by the heterodyne method (see p. 231), or the magnitude of the frequency shift may be found by measuring the capacitance needed to re-tune the oscillator to its original frequency; the latter method is used in the E. H. Sargent "Oscillometer", and the former in West's instrument, which has already been mentioned (p. 231, and Fig. V, 13). The circuit arrangements of this type of impedimeter are rather more complex than those of type 1 or 2, and the mechanical lay-out and design must be carefully planned to prevent interaction of the tuned circuits; the stability and sensitivity of a soundly constructed instrument of this type exceed, however, those of either of the other two types.

(e) *Commercial Instruments*

(i) *The "Oscillometer"*

The E. H. Sargent and Co. "Oscillometer" is a well designed instrument of Type 3 with excellent performance. It has a single oscillator, an amplifier stage, a frequency discriminator circuit, and a capacitor re-tune measuring device (Fig. V, 17). The cell is connected in parallel with the tank circuit of the oscillator, whose output is amplified and fed into the discriminator, which is permanently tuned to 5 Mc/sec. The discriminator is so designed that its voltage output is essentially zero when the oscillator frequency corresponds exactly to that of the tuned discriminator. The voltage output of the discriminator is direction-sensitive and rises sharply when the oscillator frequency deviates from the reference frequency of 5 Mc/sec. When the frequency of the oscillator alters because of a change in composition of the chemical system, the deviation is indicated by the valve voltmeter, and the oscillator is re-tuned to the reference frequency by means of a variable precision condenser in parallel with the cell; the number of scale divisions of capacitance necessary to re-tune the oscillator is used as the sensible indicator. The details of the circuit and the method of operation are adequately described in the maker's manual and need not be included here; it is, however, pertinent to point out that the instrument can be used in conjuction

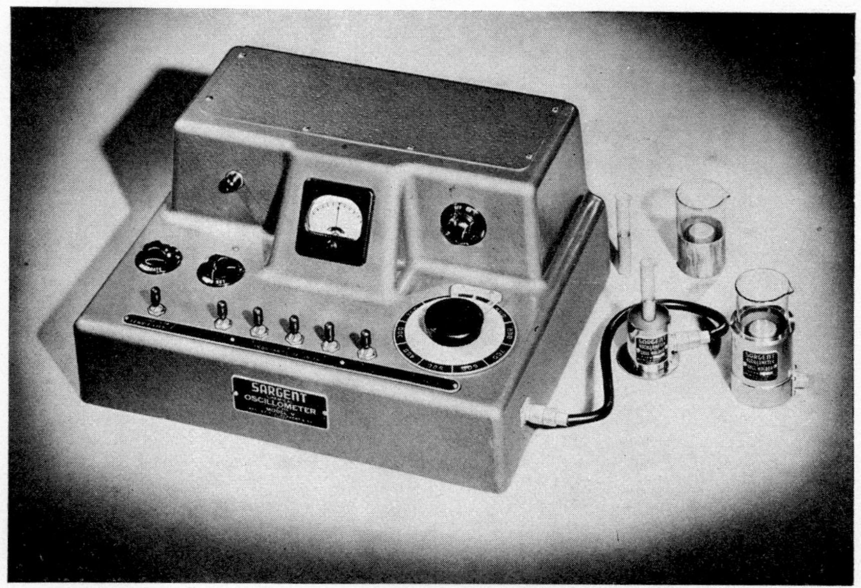

Fig. V, 17. Sargent Oscillometer Model V.
(Courtesy of E. H. Sargent and Co., Chicago, Ill.)

with the standard types of industrial potentiometric recorders: the voltage output of the discriminator circuit is impressed across a voltage divider circuit that feeds the recorder; the makers state that there is an approximately linear relationship between discriminator voltage and frequency deviation over a range of 1000 scale divisions on either side of the reference frequency.

(*ii*) *The Sargent-Jensen Titrator*

The E. H. Sargent and Co.'s Sargent-Jensen Titrator is a Type I instrument, and consists of a five-megacycle oscillator, which, through a buffer-amplifier stage, loads the primary of a transformer that is tuned to the oscillator frequency (Fig. V, 18). The condenser type of cell is part of the secondary of the transformer, and any change in the composition of the chemical system in the cell detunes the transformer secondary and thus causes a change in the voltage developed across the secondary; this voltage is rectified and measured with a valve voltmeter circuit. The titrator was primarily designed for increased sensitivity in the concentration range equivalent in

References p. 254

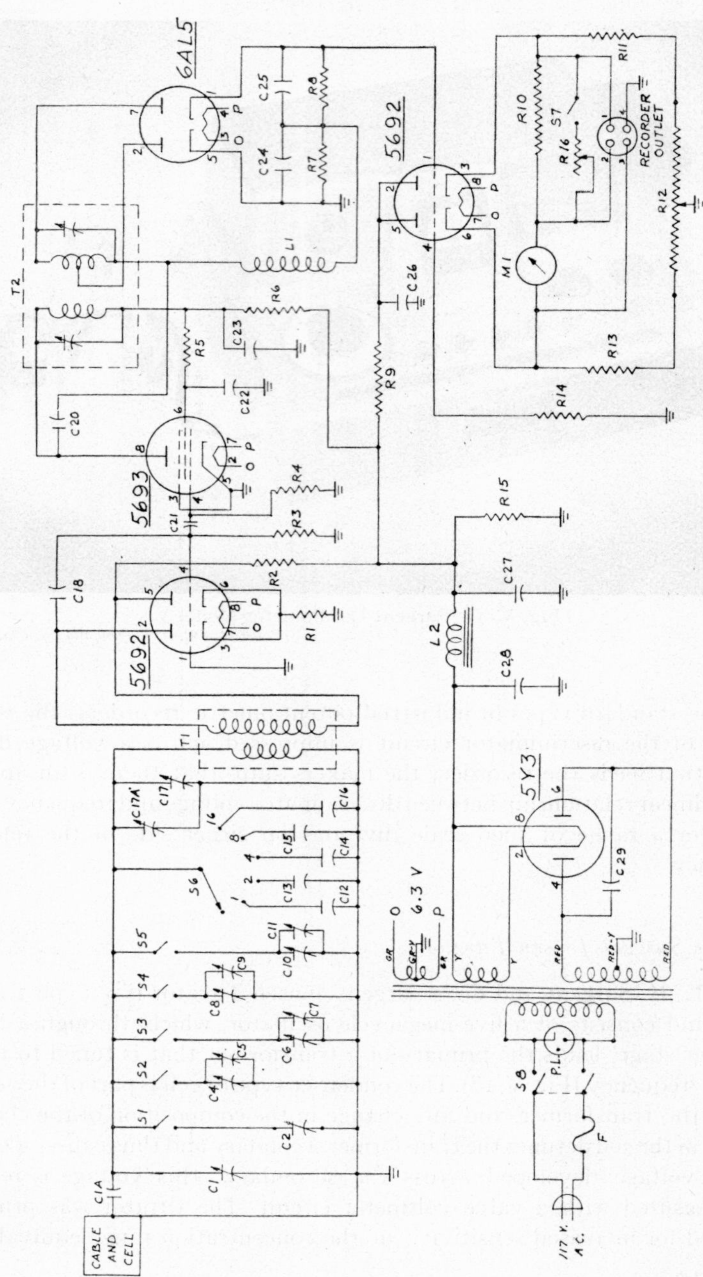

Fig. 17a. Circuit diagram of Sargent Oscillometer Model V.

Fig. V, 18. Sargent-Jensen Titrator.
(Courtesy of E. H. Sargent and Co., Chicago, Ill.)

conductivity to a $0.1N$ solution of potassium chloride, in which range its performance and stability are decidedly superior to those of the "Oscillometer"; its use is not, however, limited to this range.

(*iii*) *The High Frequency Titrimeter*

The High Frequency Titrimeter, made by the Fisher Scientific Co., is a combination of Types 1 and 2. It is designed around a tuned-plate tuned-grid oscillator with a quartz crystal in the grid circuit (Fig. V, 19a). It incorporates a capacitance re-tune measuring circuit in the plate tank with a valve voltmeter for measuring grid voltage; the two measurements indicate the magnitude of the reactive and the resistive components of the impedance of the cell system, and their availability is a very useful feature of the instrument. The shadow angle of the 6E5 cathode ray tube is used as the indicator of the reference point for the capacitance re-tune circuit. The plate tank circuit is first tuned to a frequency so far below the natural resonance frequency of the crystal that the circuit cannot oscillate; by means of the variable condenser C_3, the plate tank circuit is then gradually tuned to higher frequencies until the tube begins to oscillate. The point at which oscil-

References p. 254

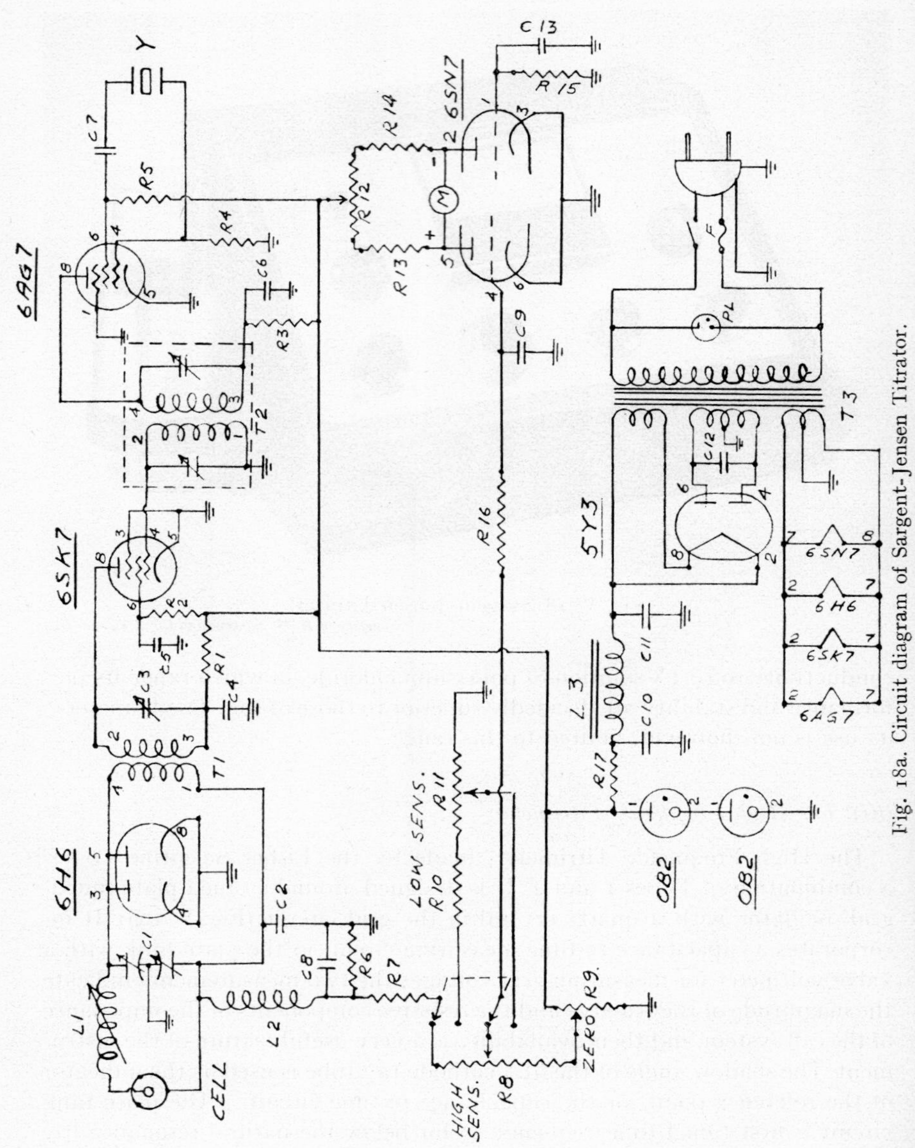

Fig. 18a. Circuit diagram of Sargent-Jensen Titrator.

Fig. V, 19. High frequency titrimeter.
(Courtesy of Fisher Scientific Co., Pittsburgh, Penn.)

lation begins is very sharp, and is indicated by a sudden decrease in the shadow angle of the 6E5 tube. The dial reading of C_3 is taken as the reference point. The difference in the dial numbers at which the shadow angle narrows is proportional to the reactance differences of the cell system.

The maximum conductivity for reactance measurements corresponds to a 0.01*N* solution of potassium chloride, while the maximum for resistive measurements corresponds to a 0.04*N* solution.

References p. 254

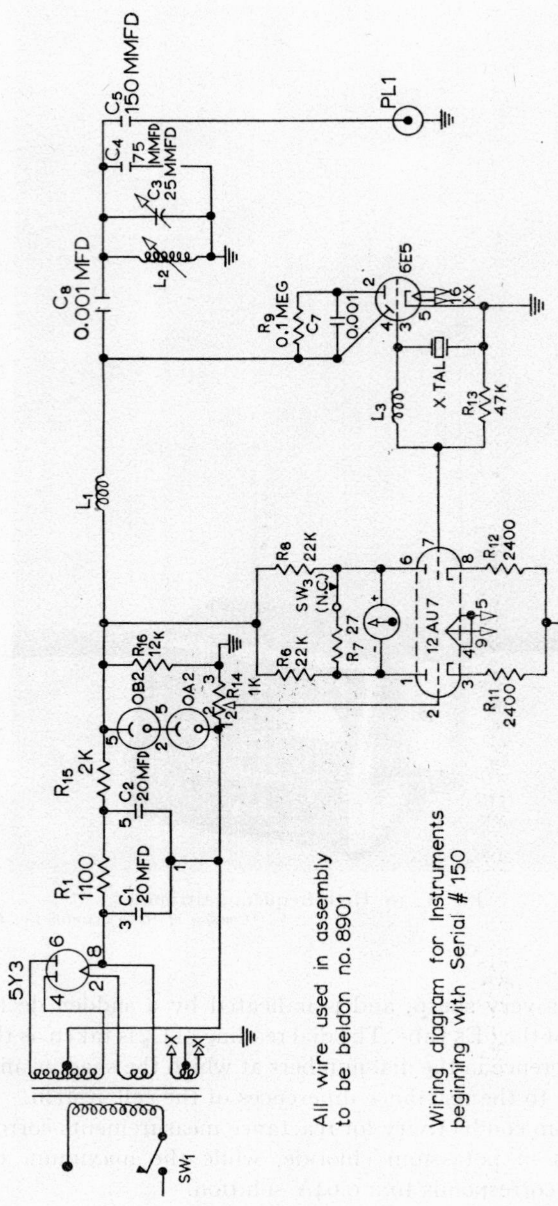

Fig. 19a. Circuit diagram of Fisher Scientific Co's High Frequency Titrimeter.

5. Applications

(a) Common Operations

The actual method of carrying out an impedimetric titration is identical with that of the more conventional conductometric titration: the solution to be titrated is placed in the impedimetric cell, the titrant is added in increments and the instrument response is read and recorded after the addition of each. A plot of volume of titrant added against instrument response gives a curve with a reversal of slope, or a sharp inflection point, which corresponds to the stoichiometric end point of the titration. The interpretation of the curve is occasionally somewhat complex, because the instrument response depends upon the conductivity of the cell system (p. 221), which does not necessarily vary linearly with the conductivity of the chemical system. In addition to titrations, instrument response may be used analytically, as though it were an indication of some fundamental physical property of the chemical system, in the same way that conductivity, molar refraction, or density are used to indicate the composition of a binary mixture; for many binary mixtures, a plot of instrument response against composition of the system yields a straight line whose slope is steep enough to be useful in determining the composition of such mixtures.

(b) Titrations

When an impedimeter is loaded by a chemical system, any change in the composition of the system causes a corresponding change in the instrument response; the instrument response can be related to the conductivity and/or the dielectric constant of the system and this relationship is the basis upon which impedimetric titrations are carried out. Within certain limits, instrument response will indicate the change in conductivity or dielectric property of the system during the course of a titration. If the initial concentration of the system falls on the sensitive portion of the instrument response curve (Fig. V, 6), an increase in conductivity of the system causes a corresponding increase in instrument response. The instrument response is thus used to follow changes in conductivity of the chemical system, and so to indicate the equivalence point of the titration. Fig. V, 20 illustrates the similarity of impedimetric and conductometric titration curves.

In considering practical applications of impedimetric titrations, it should be remembered that impedimetry is no more than a physical means of locating the end point of the titration, and offers no advantage over the use of suitable visual indicators beyond the fact that it eliminates the personal error in visual observations of colour change. Although a wide variety of neutralisation and precipitation titrations involving simple inorganic ions can be satisfactorily carried out impedimetrically, many of them can be carried out equally well and much more simply by the use of suitable visual indicators. Impedimetric titrimetry finds its most useful and practical applications when satisfactory visual indicators are not available. It should also be remembered that the means of detecting the equivalence point of a titration in no way alters the chemical properties of the titrant or of the chemical system; in general, conditions must be adjusted and interfering substances removed in exactly the same way as they would be if the titration were carried out with a visual indicator. A fundamental limitation of the method is the loss of sensitivity at conductivities greater than that corresponding to $G_{p(\text{peak})}$; in this sense, any excess of an extraneous ion constitutes an

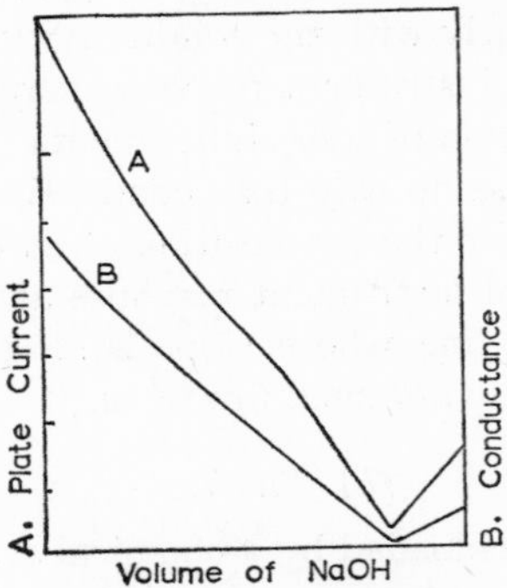

Fig. V, 20. Titration of HCl by impedimetric and conductance methods. A, impedimetric. B, conductance.

interference. In the following discussion of practical titrimetric analyses, details of sample preparation and of the suppression of interfering substances are omitted, except where the use of the impedimetric method dictates a change from the conventional procedure. No attempt has been made to list all the impedimetric titrations that have been described in the literature; instead, only those that appear to be of considerable practical use are considered.

(*i*) *Calcium*

Calcium oxide[24] in a basic metallurgical slag can be titrated directly with ammonium oxalate in the presence of precipitates of hydrated silica, ferric oxide, alumina, and other metal oxides. The slag is dissolved in hydrochloric acid, the solution is brought to pH 6 by the addition of ammonium hydroxide, and is titrated with 0.1*N* ammonium oxalate. An accuracy of $\pm$ 1% of the amount of calcium oxide present is attainable with slags containing as much as 50% of calcium oxide. Calcium hardness[25] in water can be determined by impedimetric titration with a 0.1N solution of sodium oleate that has been standardised by impedimetric titration with hydrochloric acid. Standardisation values can be reproduced to within 2%, which is better than the stated precision of the *Standard Methods for the Examination of Water and Sewage*, and the authors find that 1000 parts per million of calcium hardness can be determined with an accuracy of ± 3% of the amount present.

(*ii*) *Beryllium*

Beryllium[26] in a solution of hydrochloric acid that is free from carbon dioxide can be titrated with sodium hydroxide. The titration curve shows two distinct breaks: the first occurs at the point of neutralisation of the excess of acid; the second break corresponds to the complete precipitation of beryllium hydroxide; any metal ion that forms an insoluble hydroxide will interfere.

(*iii*) *Thorium*

The stoichiometry of the reaction between the thorium ion and the oxalate ion has not been completely elucidated; it is known, however, that when oxalate is added to a solution containing thorium, a compound of indefinite composition is present at the theoretical equivalence point; on the other hand, when thorium is added to an oxalate solution, the precipitate at the equivalence point is thorium oxalate. This reaction can be used for the indirect determination of thorium[27]: the solution containing the thorium is added to such a measured volume of a standard solution of oxalic acid that the oxalate ion will be in excess at the end of the reaction; this excess is then back-titrated with a standard solution of thorium nitrate in an acid medium, which must have a pH of 2, or less, if the solution contains less than one millimole of thorium.

(*iv*) *Titrations with 8-Hydroxyquinoline*

Goto[28] has described a procedure for the impedimetric titration of cupric, zinc, aluminium, and ferric ions with 8-hydroxyquinoline. The concentration of the titrant is 0.1N, and the maximum permissible concentration of the metal ions concerned is 0.02N for copper and iron, and 0.1N for zinc and aluminium; the titration is carried out in an acetic acid/sodium acetate buffer. As 8-hydroxyquinoline forms complex compounds with a large number of metal ions, any that would interfere must be removed.

(*v*) *Titrations with Ethylenediaminetetra-acetic Acid*

Ethylenediaminetetra-acetic acid (EDTA) forms complex compounds of

References p. 254

varying stability with nearly all metals except the alkali metals, and the titrimetric determination of those metals that form stable complexes has been investigated by a number of workers[29, 30, 31, 32, 33, 34]. EDTA (hereafter referred to as H_4Y) is a dibasic acid of medium strength and forms complexes with most metal ions in a ratio of 1 : 1, according to the following equation

$$2H^+ + H_2Y^{2-} + M^{2+} \rightarrow 4H^+ + (MY)^{2-} \qquad (10)$$

Disodium ethylenediaminetetra-acetic acid (Na_2H_2Y) reacts with metal ions as follows:

$$2Na^+ + H_2Y^{2-} + M^{2+} \rightarrow 2Na^+ + 2H^+ + (MY)^{2-} \qquad (11)$$

H_4Y acts as a dibasic acid and can be titrated directly with a strong base, while an equimolar mixture of H_4Y and a bivalent metal ion, when titrated directly with a strong base, acts as a tetrabasic acid. The difference in the number of available hydrogen ions derived from the free acid and from the complexed acid offers the means of performing an indirect titration of the metal ion present: a neutral solution containing the metal ion to be determined is added to a measured volume of a standard solution of H_4Y; the quantity of H_4Y present must be in excess of that required to react with all the metal ion. If the metal solution is then titrated with a standard solution of sodium hydroxide, the excess of free acid will react with two equivalents of base per mole of acid, whilst the complexed acid will react with four equivalents of base per mole. The number of moles of metal (equal to the moles of complexed acid) is then calculated from the following equation:

$$\text{number of moles of metal} = \frac{\text{equivalents of base} - 2(\text{moles of } H_4Y)}{2} \qquad (12)$$

It should be pointed out that this type of titration can be carried out equally well impedimetrically or with a visual indicator, because the actual titration is a simple neutralisation.

A buffered aqueous solution of a metal ion can be titrated directly with Na_2H_2Y and the end point detected impedimetrically (equation 11). The change in conductivity of the solution before the end point is caused by the removal of the metal ion by complex formation and by the addition of two sodium ions from the titrant. Beyond the end point, the conductivity of the solution increases sharply owing to the addition of excess of titrant. The resulting curve shows a sharp reversal at the stoichiometric end point of the titration (Fig. V, 21). Copper(II), zinc, nickel, manganese(II), cobalt, cadmium, iron(III) and aluminium can be titrated directly with Na_2H_2Y

in a sodium acetate/acetic acid buffer at pH 5; thorium can be titrated at pH 3 to 4; calcium at a pH of 7 and magnesium, barium, strontium, silver and thallium(I) between pH 8 and 10. The uranyl ion $(UO_2)^{2+}$ is unique in that it reacts with H_4Y in a ratio of two moles of metal to one of the acid,

$$2UO_2^{2+} + H_2Y^{2-} \rightarrow (UO_2)_2Y + 2H^+. \tag{13}$$

The titration of the uranyl ion is best carried out with trisodium ethylenediaminetetra-acetic acid as the reagent. Either the disodium or the tetrasodium salt may be used, but it has been found that, when the titration is carried out at pH 3.5, the trisodium salt yields a particularly sharp inflection point[33].

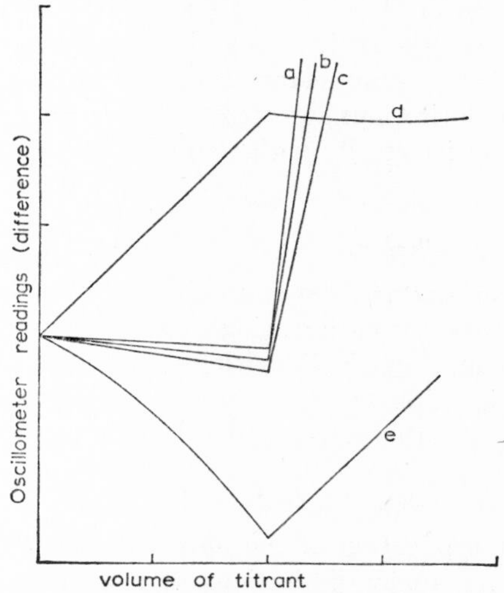

Fig. V, 21. Titration of 0.0003*M* metal salts with 0.006*M* disodium ethylenediaminetetra-acetic acid. (a) copper sulphate in acetate buffer, (b) cadmium chloride in acetate buffer, (c) cobalt chloride in acetate buffer, (d) lead nitrate without buffer, (e) calcium chloride in phosphate buffer (see reference 33).

(vi) Sulphate

The direct titration of the sulphate ion in aqueous solution with barium acetate can be effected satisfactorily[19], provided that the concentration of sulphate ion is greater than 0.0004 *M*; below this, the rate of formation of barium sulphate crystals is too slow for satisfactory titration. Milner[23] has described an excellent method for the micro determination of sulphur in crude petroleum. The sample of crude oil is oxidised by a standard combustion method to convert the sulphur to sulphate, and the solution is adjusted to pH 5.5 and freed from carbon dioxide. Ethanol is then added until the

References p. 254

solution contains 35% by volume of ethanol. A few milligrams of barium sulphate are added for seed, and the solution is titrated with 0.02 *N* barium chloride. Excellent results are obtained with samples containing from 0.01 to 1.0% of sulphur.

(*vii*) *Fluoride*

Fluoride is titrated directly with lanthanum or strontium acetate, provided that the concentration of fluoride ion is greater than 0.001*N*[19, 35]. Monand[36] has determined fluorine in organic compounds by burning the compound in a stream of oxygen and passing the effluent gases over a platinum catalyst in a fused quartz tube at 1000°, where the fluorine is converted to silicon tetrafluoride. The latter (SiF_4) is absorbed in a sodium acetate/acetic acid buffer solution and titrated with lanthanum acetate or strontium acetate. Although lanthanum acetate is said to give a sharper end point, chloride and hypochlorite ions interfere; on the other hand, the titration can be carried out satisfactorily with strontium acetate in the presence of these ions.

(*viii*) *Froth-Flotation Collectors*

Potassium ethylxanthate and sodium oleate are used as collectors in the froth-flotation process for concentrating sulphide minerals, and T. Oyama[37] has titrated them impedimetrically with no separation of the flotation mixture; he used cupric sulphate as the titrant for potassium ethylxanthate, and barium chloride for the sodium oleate.

(*ix*) *Titrations in Non-aqueous Solvents*

A very practical application of the impedimetric method is titration in non-aqueous solvents, major limitations of which have been the lack of suitable visual indicators and the absence of adequate electrode systems for potentiometric methods. The pioneer work of Jensen[4] and Wagner[38], who demonstrated the utility of impedimetric titrations in non-aqueous media, has, however, appreciably reduced these limitations. The low conductivity of most non-aqueous systems makes them particularly suitable for impedimetric measurements, because it increases the useful concentration range without overloading the oscillator. Although the impedimetric method is obviously applicable to any type of reaction involving an appreciable conductance change at the equivalence point, only non-aqueous titrations involving acid-base reactions have been described in the literature. The theory of acid-base reactions in non-aqueous solvents is adequately covered in the many reference texts on the subject and will not be reviewed here; suffice it to say that the acidic (or basic) character of a solute depends upon the relative affinity of solute and solvent for protons, and that the *strength* as an *electrolyte* of such solutions is characterised by the degree of dissociation, and depends upon both solute and solvent, as well as upon the dielectric constant of the solvent. The success of an impedimetric acid-base titration in a non-aqueous solvent must, therefore, depend upon the relative degree of dissociation of the acid or base form and upon the polarity of the solvent. The solubility of the reaction product of the neutralisation also plays an important role in determining the shape of the titration curve, because the removal of ion pairs forming a precipitate would further decrease the conductivity of the solution.

Fig. V, 22 illustrates the effects of the degree of dissociation and of the solubility of the solute upon the shape of the titration curve. The four

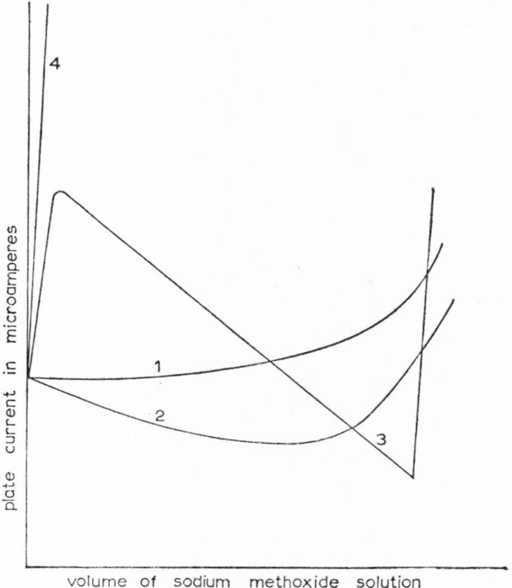

Fig. V, 22. Titration of lauric acid in various solvents with sodium methoxide in benzene. (1) Lauric acid in benzene, (2) lauric acid in benzene-acetone, (3) lauric acid in acetone, (4) lauric acid in methanol.

curves represent the titration of lauric acid in different solvents with sodium methoxide in benzene,

$$NaOCH_3 + HC_{12}H_{23}O_2 \rightarrow NaC_{12}H_{23}O_2 + CH_3OH. \quad (14)$$

Lauric acid in benzene (Fig. V, 22, curve 1) is only very slightly dissociated

References p. 254

and sodium laurate is insoluble in this solvent; therefore the only change that occurs before the end point and would effect the loading of the oscillator is the formation of CH_3OH in the solvent. Methanol, being more polar than the solvent, causes a slight loading of the oscillator, which is seen in the ascending portion of the curve before the end point. After the end point, the addition of excess of sodium methoxide, which is very weakly dissociated in benzene, causes a further increase in the loading of the oscillator, but there is no distinguishable break in the curve. Lauric acid exhibits slightly more dissociation in a benzene-acetone mixture (Fig. V, 22, curve 2) than in the benzene alone, but the sodium laurate is virtually insoluble in this mixture. The descending portion of this curve is due to the removal of the slightly dissociated lauric acid by the formation of insoluble sodium laurate, and the ascending portion is caused by the addition of excess of sodium methoxide, which is more dissociated in the benzene-acetone solvent than in benzene alone; the additional degree of dissociation of sodium methoxide results in a steeper slope over this portion of the curve than over the corresponding portion of curve 1; there is, however, still considerable curvature in the vicinity of the end point. Lauric acid in acetone (Fig. V, 22, curve 3) shows a considerable degree of dissociation; in addition, sodium laurate is slightly soluble in acetone. The first ascending portion of this curve is caused by the formation of highly ionised and soluble sodium laurate, which heavily loads the oscillator. At the maximum of this part of the curve, the solubility of sodium laurate is exceeded, and the oscillator load is reduced by the precipitation of sodium laurate, removing from the solution the partially dissociated lauric acid. At the stoichiometric end point, the addition of excess of sodium methoxide, which is highly dissociated in acetone, causes a sudden and heavy loading of the oscillator, as is indicated by the extremely steep second ascending portion of the curve. The end point is very sharp and well defined. Lauric acid in methanol (Fig. V, 22, curve 4) exhibits considerable dissociation, and the sodium laurate is quite soluble and highly ionised. The oscillator is continuously loaded during the course of the titration, first by the formation of the soluble sodium laurate, and later by the addition of excess of sodium methoxide; hence, the curve yields no indication of an end point.

Glacial acetic acid has been used extensively as a solvent for the titration of weak bases with perchloric acid, and ethylenediamine and dimethylformamide are common solvents for the titration of weak acids with sodium methoxide in benzene. Tables V, 1 and V, 2 contain a representative list of the types of acid-base titrations that have been carried out impedimetrically. The titration of the quaternary ammonium halides that are included requires the addition of enough mercuric acetate to ensure complete conversion of the

TABLE V, 1

TITRATION OF WEAK ACIDS IN VARIOUS SOLVENTS WITH SODIUM METHOXIDE

acid	*solvent*	*reference*
Salicyclic	dimethylformamide	41
Benzoic	dimethylformamide (acetone)	41, 4
Boric	dimethylformamide (ethylene glycol)	41, 42
Ammonium ion	dimethylformamide	41
Malonic	ethylene glycol	42
Succinic	ethylene glycol	42
o-Phthalic	acetone	4
Phenol	benzene-methanol (2 to 1)	44
o-Nitrophenol	dimethylformamide	41
p-Nitrophenol	ethylenediamine	40
m-Cresol	benzene-methanol (2 to 1)	44
1-Naphthol	benzene-methanol (2 to 1)	44
2-Retenol	benzene-methanol (2 to 1)	44
Pyrocatechol	benzene-methanol (2 to 1)	44
Hydroquinone	benzene-methanol (2 to 1)	44
Pyrogallol	benzene-methanol (2 to 1)	44
Phloroglucinol	benzene-methanol (2 to 1)	44
Dimedone	ethylenediamine	40
8-Hydroxyquinoline	ethylenediamine	40
2-(2-Hydroxyphenyl)benzoxazole	ethylenediamine	40
2, 3-Dihydroxy-5-methoxyquinoxaline	ethylenediamine	40
5, 8-Dihydroxy-2-isopropylquinoxaline	ethylenediamine	40
Tri-(*o*-hydroxyphenyl)phosphine oxide	ethylenediamine	40

halides to acetates. There is no exact agreement among the various investigators about the minimum acid (or base) strength required for satisfactory results in these systems: a pK_a value of 7 (in water) appears to be the minimum acid strength that gives a distinguishable break at the end point in dimethylformamide, unless one of the titration products is removed from the system by precipitation; in ethylenediamine, acids with a pK_a value of 10 (in water) have been successfully titrated; with perchloric acid in glacial acetic acid, the lower limit of base strength is a pK_a value of about 12.

Certain amino acids (I-leucine, phenylalanine and D-alanine) have been titrated[39] with perchloric acid in glacial acetic acid. The compounds were dissolved in a methanol/benzene/acetic acid mixture.

Reagents for non-aqueous titrations

Technical grade dimethylformamide is used without purification. The presence of approximately 0.03 milliequivalents of acid impurity per 100 ml necessitates the neutralisation of the solvent, or a blank correction.

References p. 254

TABLE V, 2

TITRATION OF ORGANIC BASES IN GLACIAL ACETIC ACID WITH PERCHLORIC ACID

bases	*reference*
8-Hydroxyquinoline	40
o-Phenanthroline	40
Benzidine	40
Phenazine	40
Pyridine	38, 40
2-(2-Hydroxyphenyl)benziminazole	40
Dianisyldimethylphosphonium iodide	40
Dianisylmethylphosphine	40
Tetraethylammonium bromide	40
Tetraethylammonium iodide	40
Decamethylenebispyridinium nitrite	40
Tri-(*o*-hydroxyphenyl)sulphonium chloride	40
Aniline	38, 43
p-Nitroaniline	40, 43
m-Nitroaniline	43
p-Toluidine	38, 43
p-Chloroaniline	43
p-Aminoacetophenone	43
p-Aminobenzoic acid	43
p-Anisidine	43

Glacial acetic acid to the specification of the American Chemical Society (ACS), is suitable.

Reagent grade ethylenediamine is distilled over solid sodium hydroxide.

0.1 N Perchloric acid. 8.5 ml of 70% $HClO_4$ to ACS specification is diluted to one litre with glacial acetic acid, and standardised against potassium acid phthalate; the end point is determined impedimetrically, or with crystal violet as a visual indicator.

0.1 N Sodium methoxide. 2.5 g of metallic sodium is cleaned by immersion in methanol and dissolved in 100 ml of absolute methanol. When the reaction is complete, 200 ml of methanol is added and the mixture diluted to one litre with benzene. The solution must be stored in such a way that water and carbon dioxide are excluded; it is standardised against benzoic acid in methanol with thymol blue as visual indicator.

Mercuric acetate reagent. 6 g of mercuric acetate is dissolved in 100 ml of hot glacial acetic acid and cooled to room temperature.

6. General Bibliography

K. Anderson, E. S. Bettis, and D. Revinson, *Analyt. Chem.*, 1950, **22**, 743.
K. Anderson and D. Revinson, *Analyt. Chem.*, 1950, **22**, 1272.
R. Arditti and P. Heitzmann, *Compt. rend.*, 1949, **229**, 44.

A. J. ARVIA and P. H. BRODERSEN, *Z. phys. Chem. (Frankfurt)*, 1956, **6**, 381.
T. ASADA, *J. Electrochem. Soc. Japan*, 1953, **21**, 158, 221; 1954, **22**, 16, 126, 171, 375; 1955, **23**, 30, 235.
F. BAUMANN and W. J. BLAEDEL, *Analyt. Chem.*, 1956, **28, 2.**
R. J. BEAVER, C. E. CROUTHAMEL, and H. DIEHL, *Iowa State College J. Sci.*, 1949, **23**, 289.
G. S. BEIN, *Analyt. Chem.*, 1954, **26**, 209.
P. J. BENDER, *J. Chem. Educ.*, 1946, **23**, 179.
C. BERTOGLIO and E. MARCON, *Ann. Chim. (Italy)*, 1956, **46**, 1121.
W. J. BLAEDEL, T. S. BURKHALTER, D. G. FLOM, G. HARE, and F. W. JENSEN, *Analyt. Chem.*, 1952, **24**, 198.
W. J. BLAEDEL and H. T. KNIGHT, *Analyt. Chem.*, 1954, **26**, 743.
W. J. BLAEDEL and H. V. MALMSTADT, *Analyt. Chem.*, 1950, **22**, 743, 1410, 1413; 1951, **23**, 471; 1952, **24**, 450, 455.
W. J. BLAEDEL, H. V. MALMSTADT, D. L. PETITJEAN, and W. K. ANDERSON, *Analyt. Chem.*, 1950, **22**, 1240.
G. G. BLAKE, *J. Roy. Soc. Arts*, 1933, **82**, 154.
Idem, Analyst, 1950, **75**, 32, 689; 1951, **76**, 241; 1954, **79**, 108.
Idem, Analyt. Chim. Acta, 1955, **13**, 393; 1956, **15**, 232.
Idem, Austral. J. Sci., 1947, **10**, 80; 1948, **11**, 59; 1949, **12**, 32; 1951, **14**, 93.
Idem, Chem. and Ind., 1946, (3), 28; 1947, 536; 1949, 741; 1950, 67; 1951, 59.
Idem, J. Sci. Instr., 1945, **22**, 174; 1947, **24**, 77, 101.
Idem, Conductimetric Analysis at Radio Frequency (Chapman and Hall, London, 1950; Chemical Publishing Co., New York, 1952).
C. M. BURRELL, T. G. MAJURY, and H. W. MELVILLE, *Proc. Roy. Soc.*, 1951, **A 205**, 309.
J. W. BROADHURST, *J. Sci. Instr.*, 1944, **21**, 108.
C. F. BROCKELSBY, *J. Sci. Instr.*, 1945, **22**, 243.
J. C. CLAYTON, J. F. HAZEL, W. M. MCNABB, and G. L. SCHNABLE, *Analyt. Chim. Acta*, 1955, **13**, 487.
K. CRUSE and R. HUBER, *Angew. Chem.*, 1954, **66**, 625.
K. CRUSE, *Arch. Eisenhüttenw.*, 1954, **25**, 563.
J. A. DEAN and C. CAIN, JR., *Analyt. Chem.*, 1955, **27**, 212.
P. DEBYE and H. FALKENHAGEN, *Phys. Z.*, 1928, **29**, 401.
P. DELAHAY, *Instrumental Analysis* (The Macmillan Co., New York, 1957), p. 151.
J. P. DOWDALL, D. V. SINKINSON and H. STRETCH, *Analyst*, 1955, **80**, 491.
H. DUBOIS, V. C. DAVIS, and S. W. NICKSIC, *Amer. Chem. Soc., Div. Petroleum Chem., Symposium No. 27*, 1953, 85.
F. R. DUKE, R. D. BEAVER, and H. DIEHL, *Iowa State College J. Sci.*, 1949, **23**, 297.
G. DUYCHAERTS, *Ind. chim. belge*, 1953, **18**, 795.
P. O. ENGELDER, *Instruments*, 1951, **24**, 335.
P. J. ELVING, *Discuss. Faraday Soc.*, 1954, No. 17, 156.
P. J. ELVING and J. LAKRITZ, *J. Amer. Chem. Soc.*, 1955, **77**, 3217.
R. B. FISCHER, *Analyt. Chem.*, 1947, **19**, 835.
R. B. FISCHER and D. J. FISCHER, *Proc. Indiana Acad. Sci.*, 1953, **62**, 160.
D. G. FLOM and P. J. ELVING, *Analyt. Chem.*, 1953, **25**, 541.
J. FOREMAN and D. CRISP, *Trans. Faraday Soc.*, 1946, **42A**, 186.
J. S. FRITZ, *Acid-Base Titrations in Non-Aqueous Solvents* (G. Frederick Smith Chemical Co., Columbus, Ohio, 1952).
S. FUJIWARA and S. HAYASHI, *Analyt. Chem.*, 1954, **26**, 239.
A. M. GORRIZ and F. M. MASIP, *Anales real Soc. espan. Fis. Quim.*, 1954, **50B**, 909.
H. GOTO and T. HIRAYAMA, *J. Chem. Soc. Japan*, 1952, **73**, 656.
H. GOTO, T. HIRAYAMA, and S. SUZIKI, *J. Chem. Soc. Japan*, 1953, **74**, 216.
H. GOTO and T. HIRAYAMA, *Sci. Rep. Res. Inst. Tŏhoku Univ., Ser. A*, 1954, **6**, 228.
C. L. GRANT and H. M. HAENDLER, *Analyt. Chem.*, 1956, **28**, 415.
J. J. LINGANE, *Electroanalytical Chemistry* (Interscience Publishers, New York, 1953).
E. C. MASKINER, *J. Sci. Instr.*, 1947, **24**, 219.

References p. 254

M. Masui, *J. Pharm. Soc. Japan*, 1953, **73**, 921, 1011; 1954, **74**, 530; 1955, **75**, 1519; 1956, **76**, 1109.

M. Masui and Y. Kamura, *J. Pharm. Soc. Japan*, 1951, **71**, 705, 1194, 1294.

O. I. Milner, *Analyt. Chem.*, 1952, **24**, 1247.

P. H. Monaghan, P. B. Moseley, T. S. Burkhalter, and O. A. Nance, *Analyt. Chem.*, 1952, **24**, 193.

P. Monand, *Bull. Soc. chim. France*, 1956, 704.

R. Mori, T. Hyodo, and Y. Murakami, *Rep. Res., Lab., Asahi Glass Co. Ltd.*, 1950, **1**, 76.

M. Munemori, *Bull. Naniwa Univ.*, 1954, **2A**, 97.

S. Musha, *Sci. Rep. Tôhoku Univ.*, 1951, **3A**, No. 1, 55; 1952, **4A**, 575.

Idem, Nippon Kinzoku Gakkaishi, 1950, **B14**, No. 5, 70.

S. Musha, M. Ito and M. Takeda, *J. Chem. Soc. Japan*, 1952, **73**, 482.

N. Nakamura, *Japan Analyst*, 1955, **4**, 345.

K. Nakano, *J. Chem. Soc. Japan*, 1953, **74**, 227, 345; 1954, **75**, 494, 773, 776.

K. Nakano, R. Hara, and K. Yashiro, *Analyt. Chem.*, 1954, **26**, 636.

O. A. Nance, T. S. Burkhalter, and P. H. Monaghan, *Analyt. Chem.*, 1952, **24**, 214.

L. G. Groves and J. King, *J. Soc. Chem. Ind.*, 1946, **65**, 320.

Idem, Analyst, 1948, **73**, 217.

J. L. Hall, *Analyt. Chem.*, 1952, **24**, 1236, 1244.

J. L. Hall and J. A. Gibson, Jr., *Analyt. Chem.*, 1951, **23**, 966.

J. L. Hall, J. A. Gibson, Jr., H. O. Phillips, and F. E. Critchfield, *Analyt. Chem.*, 1954, **26**, 1539.

Idem, J. Chem. Educ., 1954, **31**, 54.

J. L. Hall, J. A. Gibson, Jr., H. O. Phillips, and P. R. Wilkinson, *Analyt. Chem.*, 1955, **27**, 1504.

J. L. Hall, J. A. Gibson, Jr., F. E. Critchfield, H. O. Phillips, and C. B. Seibert, *Analyt. Chem.*, 1954, **26**, 835.

H. W. Hamme, E. L. Grove, and J. L. Kassner, *J. Chem. Phys.*, 1954, **22**, 944.

R. Hara, *J. Pharm. Soc. Japan*, 1951, **71**, 1122 [in English].

R. Hara and P. W. West, *Analyt. Chim. Acta*, 1954, **11**, 264; 1955, **12**, 72, 285; 1955, **13**, 189; 1956, **14**, 280.

J. H. Harley and S. E. Wiberley, *Instrumental Analysis* (John Wiley, New York 1954), p. 292.

Y. Hashimato and I. Mori, *J. Pharm. Soc. Japan*, 1952, **72**, 1532.

J. R. Haskins, G. Heller, and E. Miller, *J. Chem. Phys.*, 1955, **23**, 755.

Y. Hayashi, *Proc. Fujihara Mem. Fac. Eng. Keio Univ.*, 1953, **6**, No. 21, 18.

M. Honda, *J. Chem. Soc. Japan*, 1952, **73**, 529.

M. Honda, K. Nakano and A. Satsuka, *J. Chem. Soc. Japan*, 1954, **75**, 1299.

M. Honda and H. Tandano, *Japan Analyst*, 1953, **2**, 456.

R. Huber and K. Cruse, *Z. Electrochem.*, 1954, **58**, 156.

M. Ishibashi, T. Fujinaga, and M. Mitamura, *Bull. Inst. Chem. Res., Kyoto Univ.*, 1951, **25**, 24.

Idem, Japan Analyst, 1953, **2**, 195.

M. Ishidate and Y. Kamura, *J. Pharm. Soc. Japan*, 1954, **74**, 827; 1955, **75**, 260.

M. Ishidate and M. Masui, *J. Pharm. Soc. Japan*, 1953, **73**, 487.

Idem, Pharm. Bull. (Japan), 1954, **2**, 50.

F. W. Jensen, *U. S. pat.* 2,645,563, July 1953.

F. W. Jensen and A. L. Parrack, *Bull. Texas Eng. Exp. Sta.*, 1946, No. 92.

Idem, Analyt. Chem., 1946, **18**, 595.

F. W. Jensen, G. M. Watson and L. G. Vela, *Analyt. Chem.*, 1951, **23**, 1327.

F. W. Jensen, G. M. Watson, and J. B. Beckham, *Analyt. Chem.*, 1951, **23**, 1327.

F. W. Jensen, M. J. Kelly, and M. B. Burton, Jr., *Analyt. Chem.*, 1954, **26**, 1716.

A. H. Johnson and A. Timnick, *Analyt. Chem.*, 1956, **28**, 889.

Y. Kamura, *Pharm. Bull. (Japan)*, 1955, **3**, 138.

Idem, Kagaku No Ryoiki, 1952, **6**, 771.

Idem, *J. Pharm. Soc. Japan*, 1954, **74**, 1.
O. L. KAPSTAN and V. A. TEPLYAKOV, *Zhur. analit. Khim.*, 1953, **8**, 131.
K. J. KARRMAN, and G. JOHANSSON, *Mikrochim. Acta*, 1956, 1573.
JUN KATA and Y. OYAGI, *J. Coll. Arts Sci., Chiba Univ.*, 1952, **1**, 6.
F. KLUTKE, *Dechema Monograph.*, 1951, **17**, 108.
T. KONO, *J. Agric. Chem. Soc. Japan*, 1951–52, **25**, 567; 1952, **26**, 41, 533.
T. KONO and M. ISOYA, *ibid.*, 1952, **26**, 708.
S. S. KREMEN, L. M. MATHEWS, and C. R. BORDERS, *J. Amer. Leather Chemists' Assoc.*, 1949, **44**, 459.
E. S. LANE, *Analyst*, 1955, **80**, 675.
D. E. LASKOWSKI and R. E. PUSCHER, *Analyt. Chem.*, 1952, **24**, 965.
W. T. LIPPINCOTT and A. TIMNICK, *Analyt. Chem.*, 1956, **28**, 1690.
M. NISHIGAI, H. OKABAYASHI and N. TANAKA, *Rep. Radiation Chem. Res. Inst., Tokyo Univ.*, 1950, **5**, 45.
H. NOGAMI and F. NAKAGAWA, *J. Pharm. Soc. Japan*, 1955, **75**, 1289.
H. OKABAYASHI, K. NAKANO and S. FUJIWARA, *Rep. Radiation Chem. Res. Inst., Tokyo Univ.*, 1949, **4**, 23.
M. OOSTING, *Chem. Weekblad*, 1956, **52**, 665.
T. OYAMA, J. SHIMOIIZAKA and S. USUI, *Tôhoku Kozan*, 1955, **2**, No. 2, 15.
C. N. REILLEY, Chapter in P. DELAHAY, *New Instrumental Methods in Electrochemistry* (Interscience Publishers, New York, 1954).
C. N. REILLEY and W. H. MCCURDY, JR., *Analyt. Chem.*, 1953, **25**, 86.
E. H. SARGENT and Co., *Scientific Apparatus and Methods*, 1949, **2**, 68.
Idem, (O. A. NANCE) *ibid.*, 1951, **4**, 34.
Idem, (G. A. DAWE and R. KARR) *ibid.*, 1952, **5**, 4; 1953, No. 15, 2; 1957, No. 1, 9.
H. A. SACK and B. SACK, *J. Rech. Centre Natl. Rech. Sci.* (*Paris*), 1951, No. 15, 325.
YU YU SAMITOV, *Zhur. analit. Khim.*, 1956, **11**, 149.
YU YU SAMITOV and V. M. GOROKHOVSKII, *Zhur. analit. Khim.*, 1956, **11**, 621.
P. H. SHERRICK, G. A. DAWE, R. KARR, and E. F. EWEN, *Manual of Chemical Oscillometry* (E. H. Sargent and Co., Chicago, 1954).
T. TAKAHASHI, K. KIMOTO, and T. YAMADA, *J. Chem. Soc. Japan, Ind. Chem. Sect.*, 1951, **54**, 427.
T. TAKAHASHI and K. KIMOTO, *Kagaku No Ryoiki*, 1951, **5**, 324.
N. TANAKA and M. NISHIGAI, *Rep. Inst. Sci. Technol., Univ. Tokyo*, 1952, **6**, 131.
F. E. TERMAN, *Radio Engineers' Handbook* (McGraw-Hill, New York, 1943).
J. C. TRANTHAM and R. B. BEVAN, *Analyt. Chem.*, 1952, **24**, 2018.
G. V. TROITSKII, *Biokhimiya*, 1940, **5**, 375.
W. VAN TONGEREN, *Chem. Weekblad*, 1951, **47**, 281.
W. F. WAGNER and W. B. KAUFFMAN, *Analyt. Chem.*, 1953, **25**, 538.
T. WATANABE and N. KOIZUMI, *Bull. Inst. Chem. Res., Kyoyo Univ.*, 1951, **27**, 57; 1953, **31**, 58.
R. D. WEAVER, G. C. WHITNACK and E. ST. CLAIR GANTZ, *Analyt. Chem.*, 1956, **28**, 329.
P. W. WEST, *Chem. Age*, 1953, **68**, 133.
P. W. WEST, T. S. BURKHALTER, and L. BROUSSARD, *Analyt. Chem.*, 1950, **22**, 469.
P. W. WEST, T. ROBICHAUX, and T. S. BURKHALTER, *Analyt. Chem.*, 1951, **23**, 1625.
P. W. WEST, P. SENISE, and T. S. BURKHALTER, *Analyt. Chem.*, 1952, **24**, 1250.
H. H. WILLARD, L. L. MERRITT, and J. A. DEAN, *Instrumental Methods of Analysis* (D. Van Nostrand, New York, 1951), p. 314.
T. YANO, S. MUSYA, T. WADA, and T. HINO, *Chem. Eng.* (*Japan*), 1956, **20**, 339.
T. YANO and T. WADA, *ibid.*, 426.
H. ZAHN, *Z. Physik.*, 1928, **51**, 350.
V. A. ZARINSKII and D. I. KOSHKIN, *J. Analyt. Chem. U.S.S.R.*, 1954, **9**, 29. (Eng. transl.)
Idem, *ibid.*, 1956, **10**, 101. (Eng. transl.)
Idem, *Zavodskaya Lab.*, 1956, **22**, 110.
V. A. ZARINSKII and I. R. MANDEL'BERG, *Zavodskaya Lab.*, 1956, **22**, 262.

References p. 254

REFERENCES

1. H. Zahn, *Z. Physik*, 1928, **51**, 350.
2. P. Debye and H. Falkenhagen, *Phys. Z.*, 1928, **29**, 401.
3. G. G. Blake, *J. Roy. Soc. Arts*, 1933, **82**, 154; *J. Sci. Instr.*, 1947, **24**, 101.
4. F. W. Jensen and A. L. Parrack, *Analyt. Chem.*, 1946, **18**, 595.
5. T. Asada, *J. Electrochem. Soc. Japan*, 1954, **22**, 375.
6. Y. Kamura, *Pharm. Bull.* (*Japan*), 1955, **3**, 138.
7. Y. Samitov, *Zhur. analit. Khim.*, 1956, **11**, 149.
8. J. L. Hall, *Analyt. Chem.*, 1952, **24**, 1236.
9. C. N. Reilley and W. H. McCurdy, Jr., *Analyt. Chem.*, 1953, **25**, 86.
10. J. Haskins, G. Heller, and E. Miller, *J. Chem. Phys.*, 1955, **23**, 755.
11. K. Anderson, E. Bettis, and D. Revinson, *Analyt. Chem.*, 1950, **22**, 743.
12. J. Hall and J. Gibson, *Analyt. Chem.*, 1951, **23**, 966.
13. T. Asada, *J. Electrochem. Soc. Japan*, 1955, **23**, 235.
14. F. E. Terman, *Radio Engineers' Handbook* (McGraw Hill Book Co., New York, 1943), p. 480.
15. G. Cramer, *Q. S. T.*, 1948, **32**, 42.
16. W. J. Blaedel and H. V. Malmstadt, *Analyt. Chem.*, 1950, **22**, 1413.
17. J. L. Hall, *Analyt. Chem.*, 1952, **24**, 1244.
18. F. W. Jensen, M. J. Kelly, and M. B. Burton, Jr., *Analyt. Chem.*, 1954, **26**, 1716.
19. R. J. Beaver, C. E. Crouthamel, and H. Diehl, *Iowa State J. Sci.*, 1949, **23**, 289.
20. P. W. West, T. S. Burkhalter, and L. Broussard, *Analyt. Chem.*, 1950, **22**, 469.
21. G. G. Blake, *Austral. J. Sci.*, 1947, **10**, 80.
22. R. Arditti and P. Heitzmann, *Compt. rend.*, 1949, **229**, 44.
23. O. I. Milner, *Analyt. Chem.*, 1952, **24**, 1247.
24. S. Musha, *Nippon Kinzoka Gakkaishi*, 1950, **B14**, No. 5, 70.
25. F. W. Jensen, G. W. Watson, and L. G. Vela, *Analyt. Chem.*, 1951, **23**, 1327.
26. K. Anderson and D. Revinson, *Analyt. Chem.*, 1950, **22**, 1272.
27. W. J. Blaedel and H. V. Malmstadt, *Analyt. Chem.*, 1951, **23**, 471.
28. H. Goto and T. Hirayama, *J. Chem. Soc. Japan*, 1952, **73**, 656, *Chem. Abs.*, 1953, **47**, 4785.
29. W. J. Blaedel and H. T. Knight, *Analyt. Chem.*, 1954, **26**, 743.
30. R. Hara and P. W. West, *Analyt. Chim. Acta*, 1954, **11**, 264.
31. *Idem, ibid.*, 1955, **13**, 189.
32. *Idem, ibid.*, 1955, **12**, 73.
33. *Idem, ibid.*, 1955, **12**, 285.
34. H. Nogame and F. Nakagawa, *J. Pharm. Soc. Japan*, 1955, **75**, 1289, *Chem. Abs.*, 1956, **50**, 3147.
35. J. P. Dowdall, D. V. Sinkinson and H. Stretch, *Analyst*, 1953, **25**, 538.
36. P. Monand, *Bull. Soc. Chim. France*, 1956, 704, *Chem. Abs.*, 1956, **50**, 9939.
37. T. Oyama, *Tohoku Kozan*, 1955, **2**, No. 2, 15, *Chem. Abs.*, 1957, **51**, 4867.
38. W. F. Wagner and W. B. Kauffman, *Analyt. Chem.*, 1953, **25**, 538.
39. M. Masui, *J. Pharm. Soc. Japan*, 1953, **73**, 1011, *Chem. Abs.*, 1953, **48**, 1895.
40. E. S. Lane, *Analyst*, 1955, **80**, 675.
41. J. A. Dean and C. Cain, Jr., *Analyt. Chem.*, 1955, **27**, 212.
42. M. Ishidate and M. Masui, *J. Pharm. Soc. Japan*, 1953, **73**, 487, *Chem. Abs.*, 1953, **47**, 7363.
43. W. T. Lippincott and A. Timnick, *Analyt. Chem.*, 1956, **28**, 1690.
44. K. J. Karrman and G. Johansson, *Mikrochim. Acta*, 1956, 1573.

INDEX

Page numbers in bold type indicate that full experimental details of a method are given. If an entry has several page references the particular page which contains the largest amount of discussion is printed in italics.